Springer Theses

Recognizing Outstanding Ph.D. Research

Aims and Scope

The series "Springer Theses" brings together a selection of the very best Ph.D. theses from around the world and across the physical sciences. Each volume has been selected for its scientific excellence by an internationally renowned institute or department. For greater accessibility to non-specialists, the published versions include an extended introduction, as well as a foreword by the student's supervisor explaining the special relevance of the work for the field. The series as a whole provides a valuable resource both for newcomers to the research fields, and for other scientists seeking detailed background information on special questions. Finally, it serves as an accredited documentation of the valuable contributions made by today's younger generation of scientists.

How to Apply

Theses prepared at internationally leading universities and institutes may be nominated for publication by supervisors with the endorsement of the head of department. Each such department is entitled to nominate one thesis per year. Enquiries and requests for a nomination form should be sent, together with the thesis, to Ms. Nathalie Berg (see link on right).

Eligibility Criteria

To qualify for consideration, theses should fulfil the following criteria:

- They must be written in good English.
- The topic should fall within the confines of Chemistry, Physics, Earth Sciences, Engineering and related interdisciplinary fields such as Materials, Nanoscience, Chemical Engineering, Complex Systems and Biophysics.
- The work reported in the thesis must represent a significant scientific advance.
- A maximum of 30% of the text may be a verbatim reproduction from the author's previous publications.
- They must have been examined and passed during the 12 months prior to nomination.
- The theses should have a clearly defined structure including an introduction accessible to new PhD students and scientists not expert in the relevant field.
- After acceptance, theses should be supplemented by a "Supervisor's Foreword" outlining the wider scientific relevance of the work

Indexed by zbMATH.

Jayameenakshi Venkatraman

Controlling the Effective Hamiltonian of a Driven Quantum Superconducting Circuit

Doctoral Thesis accepted by Yale University, USA

Springer

Jayameenakshi Venkatraman
Broida Hall
University of California Santa Barbara
Santa Barbara, CA, USA

ISSN 2190-5053 ISSN 2190-5061 (electronic)
Springer Theses
ISBN 978-3-031-83269-7 ISBN 978-3-031-83270-3 (eBook)
https://doi.org/10.1007/978-3-031-83270-3

This Springer imprint is published by the registered company Springer Nature Switzerland AG
The registered company address is: Gewerbestrasse 11, 6330 Cham, Switzerland

Foreword

This is an outstanding thesis dissertation reporting exciting new results in the field of driven dissipative quantum systems. The question posed by the thesis is motivated by a general physics question: can a Hamiltonian system submitted to sinusoidal driving force be represented by a time-independent Hamiltonian in a suitable reference frame? This question, which has been of practical importance in celestial mechanics in order to compute efficiently planetary dynamics, is now being revisited in its quantum-mechanical version. The results reported in this dissertation are important both for the fundamental understanding of driven dissipative systems in general, and for future developments of quantum computation. It will without doubt inspire and educate many students in the domain of quantum superconducting circuits, where quanta of microwave light interact through their coupling with the Cooper pairs of a superconductor.

Department of Applied Physics
New Haven, CT, USA

Michel Devoret

Preface

Can we observe and control the quantum manifestations of the effective static Hamiltonian of a non-linear solid-state system evolving under a fast-oscillating driving force?

We have analyzed the effective static Hamiltonian of a Josephson tunnel junction-based quantum superconducting circuit driven by a microwave sinusoidal charge excitation. In this case, this effective Hamiltonian takes the form of a Kerr oscillator submitted to a squeezing drive. We experimentally discovered pairwise simultaneous degeneracies in the spectrum of this effective Hamiltonian, which represents the dynamics of a particle in a quantum double well. What underlies these simultaneous degeneracies is the unusual destructive interference of tunnel paths in the classically forbidden region, an effect revealing a hidden symmetry of the system. Not only can these degeneracies be turned on-and-off on demand, but their number is tunable: when the detuning Δ of the drive's second subharmonic from the oscillator frequency equals an even multiple of the Kerr coefficient K, $\Delta/K = 2m$, the oscillator experiences $m + 1$ exact spectral degeneracies. Importantly, these degeneracies are robust as they are completely independent of the drive amplitude. They also lead to a drastic reduction of the incoherent well-switching rate leading to our realization of a super-protected cat qubit. Our work indicates the circumstances by which the control of parametric processes via the drive frequency can provide a practical new tool for quantum technologies. Underlying this experiment is a calculation tool to transform—beyond the rotating-wave approximation—a time-dependent Hamiltonian describing a superconducting nonlinear circuit submitted to a fast-oscillating driving force to a time-independent effective Hamiltonian governing the dynamics of our quantum double-well system.

Santa Barbara, CA, USA Jayameenakshi Venkatraman

Acknowledgments

In writing this acknowledgment, I find words alone inadequate to express the profound gratitude I feel toward all those who have contributed significantly to my PhD journey. The African saying, “It takes a village to raise a child,” resonates deeply as I reflect on the community of colleagues, family, friends, and even animals and inanimate objects that have supported me throughout this process.

Firstly, I want to express my utmost gratitude to my advisor, Michel Devoret, who recognized my need for mentorship early on and committed his valuable time to guide me. Michel’s uniquely chaotic scientific approach has taught me to appreciate the beauty in the unexpected. He has not only been understanding and supportive of my research endeavors but also encouraged me to embrace failure as a learning opportunity.

I am deeply inspired by Steve Girvin’s impeccable logic and intuition, which connect seemingly disparate problems with ease. Rob Schoelkopf’s deep concern for the practical implications of academic research and his insightful questions have been crucial in my research development. Jack Harris and Liang Jiang have put tremendous effort into presenting state-of-the-art quantum optics in a pedagogical manner, furthering my love for the field. Alexandre Blais’s enthusiasm and genuine care for the development of his younger colleagues have been invaluable in my growth as a researcher. Thanks to Francesco Iachello, Mark Dykman, Leonid Glazman, and Doug Stone for their insightful comments during research discussions and presentations.

Throughout my PhD, I have had the privilege to work alongside talented peers, such as Xu Xiao. Xu’s creativity has inspired me throughout my PhD. Facing ambitious projects with him helped me approach them with curiosity and excitement even when I was afraid. Rodrigo Cortiñas has constantly pushed my scientific beliefs and provided mentorship at a critical stage. Rodrigo’s passion and scientific nimbleness is inspiring. Yaxing Zhang sowed the seeds of the theorist in me. Yaxing is a meticulous scientist with sound scientific intuition and a great level of patience. Shruti Puri’s demanding and inspiring research provided fresh perspective to a beaten-to-death quantum optics system.

I am grateful for the support and camaraderie of the members of Qulab, Rob Schoelkopf Lab (RSL), Yale Quantum Institute (YQI), and the many others who have helped me grow both as a researcher and as a person. The friendships and connections I have formed with colleagues of all levels have enriched my experience and fueled my excitement for the future.

Thanks to the rest of Qulab, RSL, and YQI, who have helped me grow as a researcher and person: Philippe Campagne-Ibarcq, Katie Chang, Kevin Chou, Ben Chapman, Tom Connolly, Yanni Dahmani, Spencer Diamond, Andy Ding, Alec Eickbusch, Valla Fatemi, Nick Frattini, Luigi Frunzio, Suhas Ganjam, Yvonne Gao, Stijn De Graaf, Alex Grimm, Max Hays, Sumeru Hazra, Neel Thakur, James Teoh, Ioannis Tsioutsios, Takahiro Tsunoda, Pei-Kai Tsai, Katerina Kargioti, Kaicheng Li, Yao Lu, Gangqiang Liu, Alessandro Miano, Aniket Maiti, Heekun Nho, Shraddha Singh, Zlatko Minev, Gautham Uma Sankar, Kyle Serniak, Shyam Shankar, Kevin Smith, Billy Kalfus, Ben Brock, Jacob Curtis, Baptiste Royer, Baptiste Royer, Vijay Jain, Maggie Pavlovich, Alex Read, Jenny Giampalmo, Gabrielle Roberts, Yiwen Chu, Chris Wang, Zhixin Wang, Evan Zalys-Geller, Taekwan Yoon, Angela Kou, Patrick Winkel, Danny Weiss, and Neel Thakur. Shoutout to my younger colleagues, I've had a lot of fun getting to know them in my final year and am looking forward to how their research trajectories turn out: Yanhao Wang, Qile Su, Neel Thakur, Harsh Babla, Kaavya Sahay, Sophia Xue, Cassady Smith, and John Garmon. Shoumik Chowdhury and Paola Frunzio have been great mentees and I look forward to their future endeavors. I was fortunate to befriend Nuch Graves during my PhD. Nuch was always sharing her wisdom and experiences when I was taking failures too personally. I am also grateful to Terri Evangeliste, Giselle Maillet, Maria Rao, Florian Carle, Racquel Miller, Jessica, and Alex Bozzi for tirelessly working to make the fourth floors of Becton and YQI a better place to work in.

My grad school experience was made richer due to the invaluable mountaineering, climbing, and hiking adventures and following excited discussions with Rodrigo Cortiñas, Vlad Kurilovich, Pavel Kurilovich, Alec Eickbusch, and Vlad Sivak. Rodrigo has been a great friend and I will miss the numerous adventures, fights, and fun we have had together. The twins' passion was inspiring and goofiness infectious. Getting to know Christa Flühmann and Judith Höller has been a source of joy, and I cherish the long walks and thought-provoking lunches I've had with them. Thanks to Akshay for patiently answering my Field Programmable Gate Array (FPGA) and fridge-related questions when I returned to the lab after a three-year-long hiatus. Akshay, Vidul, and Ioannis have been awesome officemates, and have come to my rescue the many times I've locked myself out of my office. In my final year, Charlotte Böttcher has been a supportive colleague giving me valuable scientific advice. I will fondly remember my final year adventures with Yanhao, Qile, Neel, Kaavya, and Harsh.

Outside of the lab, I had great fun learning about plants with Rimpi Rajput and having weekend adventures with Sumita Ghosh and Apala Chaudhuri. Hiking trips with Shalini Gupta and Rohan Chitnis were adventures to always look out for. Quizzing sessions with Anushka Jha, Sarthak Chandra, Shalini Gupta, Rohan Chitnis, Sanjana Gopal, Praveen Honhar, Alankrita Bhatt, and Supranta Boruah during the pandemic helped foster a sense of connectedness and fun during a stressful period. Thanks to Shankar and Uma Shankar for warm lunch invitations throughout my PhD.

Pratyush Sarkar has been a constant source of support, encouragement, and inspiration throughout my PhD. Pratyush believed in me even when I didn't believe in myself and wanted to give up. His passion and willingness to deconstruct abstract mathematical concepts gave rise to many insightful learning opportunities. I would like to express my gratitude to Elinor for her collaboration and support in my endeavor to lead an examined life.

My parents Mangalam Venkatraman and Venkatraman Chandrasekaran have showered me with endless support and affection throughout my life. Despite having non-technical backgrounds themselves, they have made an effort to appreciate my own technical interests and encouraged me to pursue them. My heartfelt appreciation goes to my brother, Raghavendra Venkatraman, whose tenacity and kindness have been a constant source of inspiration. Alongside my sister-in-law, Vaishnavi Shankar, they have enveloped me with warmth and love throughout my graduate studies. My niece's delightful antics and boundless curiosity have been a source of joy, providing much-needed smiles on difficult days. Lastly, I wish to acknowledge my beloved cat, Miso, whose endearing personality and playful antics have brought light to even the most challenging days.

Contents

List of Symbols

$h = 2\pi\hbar$	Planck constant
e, $\Phi_0 = h/2e$	Charge quantum, magnetic flux quantum
$\hat{F}, \hat{\Box}$	Generic Hilbert-space operator (the box is a place-holder)
$F, \Box$	Generic phase-space function
$\hat{X}, \hat{P}$	Position, momentum operators of mechanical oscillator
$X_{\text{zps}}, P_{\text{zps}}$	Position and momentum coordinate zero-point spreads
M, k	Mechanical oscillator mass and spring constant
$\hat{\Phi}, \hat{Q}$	Flux, charge operators of electromagnetic oscillator
$\Phi_{\text{zps}}, Q_{\text{zps}}$	Flux, charge coordinate zero-point spreads
L, C	Electromagnetic oscillator inductance and capacitance
ω_o, Z_o	Oscillator plasma frequency, impedance
$\hat{a}, \hat{a}^\dagger$	Bosonic annihilation, creation operators
λ	Dimensionless rescaling constant of phase-space
$\hat{x}, \hat{p}$	$\hat{x} = \sqrt{\frac{\lambda}{2}}\frac{\hat{X}}{X_{\text{zps}}}$, $\hat{p} = \sqrt{\frac{\lambda}{2}}\frac{\hat{P}}{P_{\text{zps}}}$
$\hat{\varphi}, \hat{N}$	$\hat{\varphi} = \sqrt{\lambda}2\pi\frac{\hat{\Phi}}{\Phi_0}$, $\hat{N} = \sqrt{\lambda}\frac{\hat{Q}}{2e}$
S	Generating function
$\hat{U}$	Unitary operator
L_S	Lie derivative generated by S
ϵ	Parameter generating the Lie deformation
$\mathcal{T}$	Time-ordering operator
$\Box_1 \star \Box_2$	Groenewold star product between phase-space functions $\Box_1$ and $\Box_2$ (Wigner space)
$\Box_1 \star \Box_2$	Husimi star product between phase-space functions $\Box_1$ and $\Box_2$ (Husimi space)
$\{\Box_1, \Box_2\}$	Poisson bracket between phase-space functions $\Box_1$ and $\Box_2$
$\{\Box_1, \Box_2\}$	Moyal bracket between phase-space functions $\Box_1$ and $\Box_2$
$[\hat{\Box}_1, \hat{\Box}_2]$	Commutator between operators $\hat{\Box}_1$ and $\hat{\Box}_2$
$\hat{\rho}$	Density operator
W	Wigner function
$\mathfrak{W}$	Wigner transform

$\mathfrak{W}^{-1}$	Weyl transform
C	Characteristic function
K	Kerr coefficient
$\mathcal{K}$	Kamiltonian
t	Time
t_0	Initial time
E_{J}	Josephson energy
E_C	Charging energy
$\|n\rangle$	Fock state with n excitations
$\|\alpha\rangle$	Displaced Fock state or coherent state with mean photon number n
ω_a	Transition frequency between lowest Fock states in the SNAIL circuit oscillator
g_3, g_4, g_n	Oscillator third, fourth and nth rank nonlinearity
ω_d	Drive frequency
Ω_d	Drive amplitude in lab frame
Π	Drive amplitude in displaced frame
H_{eff}	Effective Hamiltonian
H_{F}	Floquet Hamiltonian
ϵ_2	Squeezing drive amplitude
δ^{ac}	Drive-power-dependent frequency dressing, or ac Stark shift
$\tilde{\omega}_a = \omega_a + \delta^{\mathrm{ac}}$	Dressed transition frequency for the lowest oscillator states
$\tilde{\square}$	Dressed $\square$
Δ	$\tilde{\omega}_a - \frac{\omega_d}{2}$
Δ^{bare}	$\omega_a - \frac{\omega_d}{2}$
Δ_o	$\omega_o - \frac{\omega_d}{2}$
j, l, m, k	Generic integers used as summation indices

List of Abbreviations

ac	Alternating Current
ac Stark shift	ac drive-power-dependent frequency renormalization
avg	AVeraGing
BCH	Baker-Campbell Hausdorff identity
BBQ	Black-Box Quantization
CT	Canonical Transformation
CQED	Cavity Quantum ElectroDynamics
cQED	circuit Quantum ElectroDynamics
DAC	Digital-Analog Converter
dc	Direct Current
eff	EFFective
EBK	Einstein-Brillouin-Keller
EOM	Equation Of Motion
exp	EXPeriment
ext	EXTernal
GKP	Gottesman-Kitaev-Preskill
LHS	Left Hand Side
max	MAXimum
min	MINimum
pr	PRobe
QEC	Quantum Error Correction
QND	Quantum Non-Demolition
rf	Radio Frequency
RHS	Right Hand Side
RWA	Rotating-Wave Approximation
sim	SIMulation
spec	SPECtroscopy
SKO	Squeeze-driven Kerr Oscillator
SNAIL	Superconducting Nonlinear Asymmetric Inductive eLement
SQUID	Superconducting QUantum Interference Device
thy	THeorY
WKB	Wentzel–Kramers–Brillouin

Chapter 1
Preamble

Abstract In this chapter, we introduce the concept of an effective Hamiltonian through two canonical examples, introduce the main questions answered in this dissertation work with relevant background both from the perspective of quantum tunneling and quantum information physics. We conclude with an outline for the rest of the dissertation document with a roadmap that highlights the logical connections between chapters.

1.1 Effective Hamiltonians in Physics

Submitting an ordinary Hamiltonian system to a rapid, time-periodic forcing field can transform it into a novel system displaying dynamical properties that have no equivalent in the class of purely static, i.e. time-independent, Hamiltonian systems. For a sufficiently perturbative forcing field (which we shorten into "drive" in the following), this novel dynamics is governed, after a proper frame transformation, by a *static effective Hamiltonian.* To be more concrete, let us take two canonical examples of effective Hamiltonians in physics.

Kapitza Pendulum

We first take as an illustration of an effective Hamiltonian system the case of a rigid pendulum of length l and mass m in the presence of gravitational acceleration g and whose pivot is submitted to a sinusoidal vibration of amplitude r along the vertical. This driven pendulum, known as the Kapitza pendulum [86, 94], is described by the classical Hamiltonian

$$H(t) = \frac{p_\theta^2}{2J} - J\omega_o^2 \cos\theta - J\frac{r}{l}\omega^2 \cos\theta \cos\omega t, \tag{1.1}$$

J. Venkatraman, *Controlling the Effective Hamiltonian of a Driven Quantum Superconducting Circuit*, Springer Theses,
https://doi.org/10.1007/978-3-031-83270-3_1

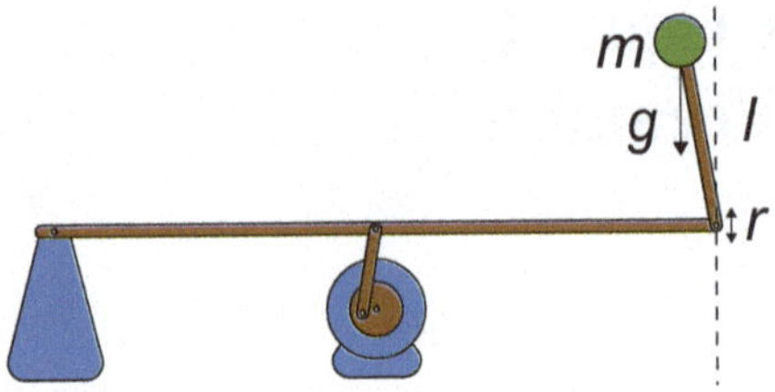

Fig. 1.1 Schematic of the Kapitza pendulum, a mechanical system which is stabilized, in the presence of a parametric drive, from a configuration which is unstable without drive

where $J = ml^2$ is the moment of inertia and $\omega_o = \sqrt{\frac{g}{l}}$ is its small oscillation frequency. In Eq. (1.1), θ is the angle between the pendulum and the vertical and p_θ is the angular momentum so that $\{\theta, p_\theta\} = 1$. The special property is that depending on the direction of these vibrations, the bob can be stabilized in new configurations unlike an ordinary rigid pendulum which can only be stabilized at the ordinary potential minimum corresponding to $\theta = 0$. For instance, when the pivot is vibrated vertically, the bob can be stabilized in an inverted configuration, at $\theta = \pi$! This gravity-defying stabilization of this system, due to and in the presence of the rapid vibration is also referred to as dynamical stabilization.

When $\omega_o \ll \omega$ and $r \ll 2l$, the transformation of the undriven system to the driven one—the dynamically stabilized one—is captured as

$$H = T(p_\theta) + V(\theta) \rightarrow H_{\text{eff}} = T(p_\theta) + V_{\text{eff}}(\theta) + \mathcal{O}\left(\frac{1}{\omega}\right), \tag{1.2}$$

$$T(p_\theta) = \frac{p_\theta^2}{2J}; \qquad V(\theta) = -J\omega_o^2 \cos\theta; \qquad V_{\text{eff}}(\theta) = V(\theta) - J\frac{r^2}{l^2}\frac{\omega^2}{8}\cos 2\theta, \tag{1.3}$$

where $T(p_\theta)$ corresponds to the kinetic energy term, $V(\theta)$ corresponds to the potential energy function in the absence of the drive, and V_{eff} corresponds to an *effective potential* energy function with stable equilibria at $\theta = 0, \pi$; see Appendix C for details on the derivation of H_{eff}. This expression for H_{eff} in Eq. (1.2) is perturbative, the perturbation parameters are the amplitude of the vibration r relative to the length of the pendulum l and the frequency of vibration ω relative to the small-oscillation frequency ω_o of the pendulum.

The magic of the effective Hamiltonian (H_{eff}) or the effective potential (V_{eff}) in this case may already have become clear to the reader: instead of resorting to complicated numerics, the transformation associated with the drive is accounted for by making a trivial replacement: $V \rightarrow V_{\text{eff}}$.[1]

[1] At higher perturbation orders, the simple separation of the effective Hamiltonian as comprising potential and kinetic energy parts does not hold as we demonstrate in Appendix C since the effective angular momentum coordinate depends both on the undriven angular momentum (p_θ) and the angle (θ).

After this mechanical system, we treat, in our next example, another important effective Hamiltonian system: the Paul trap, which plays a critical role in the field of quantum information with trapped ions.

Paul Trap

It is not possible to trap a charged particle, such as an ion, in all three directions in a static configuration of electric fields. This no-go theorem is called the Earnshaw's theorem. However, one can find a workaround to the Earnshaw theorem, to construct an average confining potential in all three directions, by submitting a trap to rapidly oscillating electric fields. In this case, the particle is still always unstable in one direction but since that direction changes with the RF drive frequency, the particle is dynamically trapped.

The Paul trap Hamiltonian can be written as

$$H(\vec{r}, \vec{p}, t) = \frac{|\vec{p}|^2}{2m} + V(\vec{r}, t), \tag{1.4}$$

where

$$V(\vec{r}, t) = \Omega \cos(\omega t)(x^2 - y^2) - \beta(x^2 + y^2 - 2z^2). \tag{1.5}$$

We have defined position and momentum coordinates of an ion-particle $\vec{r} = (x, y, z)$ and $\vec{p} = (p_x, p_y, p_z)$. The particle is located in an electrostatic saddle potential. The parameters Ω and β are electric field gradients and ω is the radio-frequency of the Paul trap. After averaging over one period, the effective Hamiltonian experienced by the particle can be written as

$$H_{\text{eff}}(\vec{r}, \vec{p}) = T(\vec{p}) + V_{\text{eff}}(\vec{r}), \tag{1.6}$$

where

$$V_{\text{eff}}(\vec{r}) = \frac{\Omega}{\sqrt{2}\omega}|\vec{r}|^2 + 2\beta z^2. \tag{1.7}$$

Remarkably, a saddle-shaped surface that is rotating and submitted to a rapidly oscillating rf field, allows a bypass of the Earnshaw theorem, i.e., by carefully choosing the Paul trap frequency ω, the origin can be transformed from an unstable point in a saddle configuration to a stable point in a harmonic trap.

1.2 Effective Hamiltonians in Quantum Superconducting Circuits

We spot two unifying features of the model systems treated so far:

First, the effective Hamiltonian's manifestations in both the Kapitza pendulum and the Paul trap are completely classical in nature.

Second, in both the Kapitza pendulum and the Paul trap, the effect of the drive is obtained, to a good approximation, by keeping the undriven kinetic energy term unchanged and by replacing the undriven potential energy term by a transformed effective potential energy, i.e.,

$$H(X, P) = \frac{P^2}{2M} + V(X) \rightarrow H_{\text{eff}}(X, P) = \frac{P^2}{2M_{\text{eff}}} + V_{\text{eff}}(X) + \mathcal{O}(1/\omega), \quad (1.8)$$

where M_{eff} denotes an effective mass of the particle, which gets renormalized due to the drive, as we discuss in Chap. 2 (see also [106, 153, 170]).

It is natural to ask:

1. Can we control the *quantum* properties of a system with an engineered effective Hamiltonian (H_{eff})?
2. Can the control brought about by the drive be exploited, for instance, by engineering effective Hamiltonians with more complicated structures than just effective potentials?

The field of circuit quantum electrodynamics (cQED) is perhaps most suited to address these questions. In circuits, Josephson tunnel junctions fulfil the wishlist of large controllable nonlinearities and weak dissipation. Moreover, state-of-the-art microwave electronics have enabled the control of individual quantum systems and their high-fidelity and quantum non demolition (QND) readout.

This thesis work answers the above questions by realizing a squeeze-driven Kerr-nonlinear oscillator (SKO) in a rapidly-driven superconducting circuit. The SKO model system that we have engineered is governed by the effective Hamiltonian:

$$\hat{H} = \Delta \hat{a}^\dagger \hat{a} - K \hat{a}^{\dagger 2} \hat{a}^2 + \epsilon_2 (\hat{a}^{\dagger 2} + \hat{a}^2). \quad (1.9)$$

Equation (1.9) corresponds to an elementary quantum system: a Kerr oscillator dressed by a squeezing interaction. Here $\hat{a}$ is the bosonic annihilation operator, and Δ, K, ϵ_2 characterize the frequency of the oscillator, its Kerr-nonlinearity and the strength of the squeezing interaction respectively.[2] These parameters are controlled in-situ by tuning the drive parameters as we discuss in Chap. 4.

Just like the effective potentials of the Kapitza pendulum and the Paul trap, the simple structure of Eq. (1.9) abstracts the fast-oscillating drive away, i.e, Eq. (1.9)

[2] The detuning parameter Δ is not to be confused with the superconducting gap.

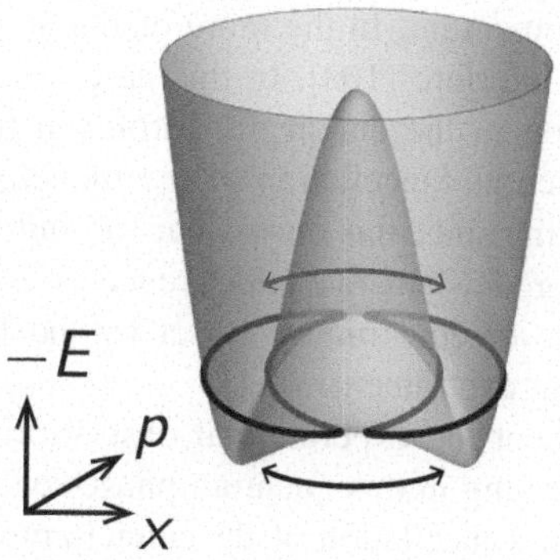

Fig. 1.2 Classical phase space surface of the effective Hamiltonian of a squeeze-driven Kerr oscillator. The parameters are taken to be $\Delta_o/K = 3$ and $\epsilon_2/K = 0.11$ in Eq. (1.10). The black curves represent Bohr-like orbits corresponding to quantized states (see Sect. 3.1). The arrows represent the two tunnel paths across classically forbidden regions in phase space

corresponds to the effective description of a superconducting circuit oscillator submitted to a rapidly-oscillating drive. However, unlike the effective potentials of the Kapitza pendulum and the Paul trap, Eq. (1.9) cannot be written down as a quadratic kinetic energy + an effective potential energy function. This can be easily seen by introducing phase space coordinate operators $\hat{a} = (\hat{x} + i\hat{p})/\sqrt{2}$ and then looking at $\hat{H}$ in classical phase space as:

$$H = \Delta_o \left(\frac{x^2 + p^2}{2}\right) - K \left(\frac{x^2 + p^2}{2}\right)^2 + \epsilon_2(x^2 - p^2), \tag{1.10}$$

Eq. (1.10) has been obtained from Eq. (1.9) following Dirac's mantra [41]: to transform an operator or a q-number $\hat{H}$ to a classical phase space function or a c-number H, (1) drop all the hats and (2) replace any non-commutative operator product by an ordinary product.[3]

Remarkably, Eq. (1.10) has a *quartic*, and not quadratic, dependence in momentum p and position x. We see this in Fig. 1.2, which shows a representation of H, in the three dimensions of position x, momentum p, and energy (E) for $\Delta/K = 3$ and $\epsilon_2/K = 0.11$, and the parameters were chosen to yield a double-well system in the phase space of x and p.

The two phase-space wells are connected via two saddle points, under which the system can tunnel. At these saddle points, the momentum is non-zero. By contrast, for a massive particle moving in an ordinary double-well potential, tunneling through the barrier is associated with only one path under the barrier maximum,

[3] We have adapted this poor man's way of transforming a classical expression to a quantum one (and vice versa) here for the sake of simplicity, but will discuss rigorously the quantum, semi-classical, and classical representations of Eq. (1.9) in Chap. 3.

corresponding to zero momentum. In the more elaborate situation of Fig. 1.2, the two tunneling paths can interfere [103]. In this case, oscillations accompany the decay of the wavefunction in the classically forbidden region. This interference can even lead to the coherent cancellation of the tunneling amplitude altogether. This is especially interesting since this may occur for finite barrier height, allowing the tunneling to be restored when the interference is constructive. Whether the interference is destructive or constructive is set by the Hamiltonian parameters, which are adjustable in our experiment.

In this thesis, we present the experimental observation of the controlled cancellation of quantum tunneling in this quantum phase-space double-well system of the SKO. We measure the cancellation of the tunnel amplitude simultaneously in the oscillator's ground and excited state manifold. This phenomenon is periodic in Δ, with the period given by $2K$ and occurs due to the destructive interference of tunneling paths in the classically forbidden region. We further identify Δ as a knob to obtain an overall exponential reduction of tunneling in the ground and excited state manifold of the SKO. These two effects contribute to the drastic enhancement of the transverse relaxation lifetime of the qubit encoded in the ground state manifold of our oscillator. The dynamical control of tunneling finds applications in quantum information processing, quantum simulation of lattice gauge theories, molecular, and nuclear physics.

Background: Quantum Tunneling

Although quantum tunneling was discovered nearly a century ago [109] and observed since in a variety of natural and synthetic systems, the treatment of tunneling is usually limited to the ground states of the system and has rarely been discussed for excited states in the literature, as we elaborate in the following survey. The phenomenology of ground state tunneling has been studied in cold atoms [132] in three-dimensional optical lattices [48], optical tweezers [87], ion traps [118] and in quantum dots [77]. In Josephson tunnel circuits, quantum tunneling of the phase variable was first observed by Devoret et al. [39] and since then exploited in several other experiments [155]. Furthermore, the tunnel effect has been involved in quantum simulation [13], in Floquet engineering of topological phases of matter and to generate artificial gauge fields with no static analog [58, 166]. The quantum interference of tunneling for the ground states of a large spin system was measured previously in a cluster of eight iron atoms by Wernsdorfer and Sessoli [158] (see also [5, 6, 100]).

Weilinga and Milburn [161] first identified that the quantum optical model in Eq. (1.9) exhibits ground state tunneling for a particular value of Δ. Marthaler and Dykman [102, 103] developed a WKB treatment for a range of the Δ parameter, and predicted that, for this model, the tunnel splitting of the ground state manifold crosses zero periodically and is accompanied by oscillation of the wavefunction in the classically forbidden region.

Our work is the first experimental realization of the longstanding theoretical proposals of the last paragraph. It is similar, but different, to the phenomenology of the "coherent destruction of tunneling", discovered theoretically by Grossmann et al. [68] and observed experimentally in cold atoms [28, 98]. Indeed, the dynamical tunneling in our experiment is in sharp contrast with photon-assisted or suppressed tunneling in weakly driven double-well potentials. Firstly, our tunneling is completely dynamical, i.e., the tunneling barrier vanishes in the absence of the drive. Secondly, and most importantly, our work extends the coherent cancellation of tunneling to all the excited states in the well. The periodic resonance condition $\Delta/K = 2m$, shared for the $m + 1$ first pairs of excited levels, is independent of the drive amplitude. Remarkably, under this multi-state resonance condition, the first $2(m + 1)$ oscillator states have a closed-form expression in the Fock basis (see supplement). We further emphasize that the dynamical tunneling in our work is distinct from chaos-assisted dynamical tunneling [149] observations made in ultracold atoms over three decades ago [75, 149]; remarkably our strongly driven nonlinear system remains integrable. To the best of our knowledge, our work corresponds to the discovery and the first demonstration of the exact simultaneous cancellation of the tunnel splitting for the ground and excited states. Our data featuring the incoherent dynamics can be qualitatively modeled by a Lindbladian treatment that we present in the supplement, yet more research on the decoherence of driven nonlinear driven systems is needed to get a quantitative agreement (see [152]).

As a resource for quantum information, the squeeze-driven Kerr oscillator for $\Delta = 0$, was identified in theory proposals by Cochrane et al. [33] and Puri et al. [126] due to its exponential resilience to low frequency noise and was proposed for a bosonic code, called the Kerr-cat qubit which we next survey. The code was implemented for the first time in circuits [66].

Background: The Kerr-cat Qubit

Ultimately performing useful quantum tasks is a key motivation to control the quantum properties of our effective Hamiltonian system. After all, the double-well system renders itself naturally to encoding a qubit, as evidenced by the flux qubit [29], the fluxonium [101], among many other solid state double-well qubits. In this section, we summarize the quantum information processing tasks that have been realized thus far with the SKO, and compare them with this thesis work.

The lowest eigenstate manifold of a double-well potential contains nearly degenerate and orthogonal, and thus naturally suggests a qubit encoding. The phenomenon of quantum tunneling under the barrier breaks the desired degeneracy though. However, when $\Delta = 0$, the ground states are exactly degenerate. This was first identified in theory proposals by Cochrane et al. [33] and Puri et al. [126]. This property can be understood by writing Eq. (1.9) for $\Delta = 0$ into the factorized form [126]

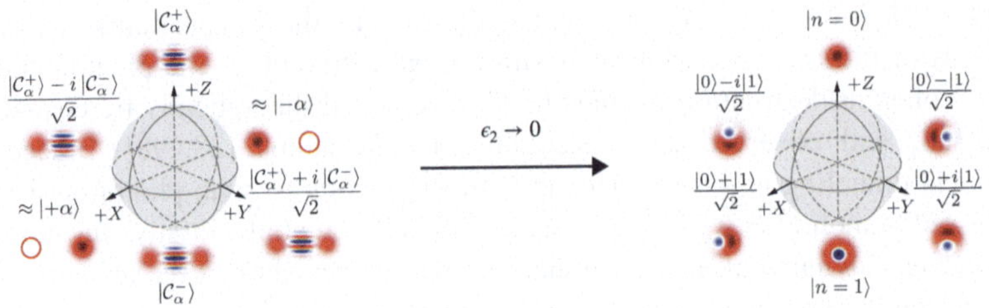

Fig. 1.3 Bloch sphere of the Kerr-cat qubit, first introduced in [66, 126]. The Bloch sphere on the left hand side (LHS), which corresponds to the encoding for non-zero ϵ_2, has the even and odd Schrödinger cat states in the north and south poles. It reduces to the Bloch sphere on the right hand side, where the lowest two Fock states form north and south poles, in the absence of the squeezing interaction

$$\hat{H}_{\Delta=0} = -K(\hat{a}^{\dagger 2} - \epsilon_2/K)(\hat{a}^2 - \epsilon_2/K), \tag{1.11}$$

from which it follows that the two coherent states $|\pm\alpha\rangle$ with $\alpha = \sqrt{\epsilon_2/K}$, which are the eigenstates of the annihilation operator $\hat{a}$, are also degenerate eigenstates of Eq. (1.11). Since Eq. (1.11) is negative-semidefinite and $\hat{H}_{\Delta=0}|\pm\alpha\rangle = 0$, these states are the ground states. This property motivated the proposal in [126] to use this degenerate manifold in the double-well system as a qubit. This bosonic encoding was implemented for the first time in quantum superconducting circuits and titled the *Kerr-cat qubit* [66].

The original Kerr-cat qubit encoding is shown in Fig. 1.3. The Bloch sphere on the left hand side (LHS) corresponds to the encoding for $\epsilon_2 \neq 0$ and it reduces to the Bloch sphere on the right hand side (RHS) corresponding to $\epsilon_2 = 0$. The even and odd Schrödinger cat states are taken to be in the north and south poles of the sphere[4] and they adiabatically map to the $|0\rangle$ and $|0\rangle$ Fock states of the Kerr oscillator as $\epsilon_2 \to 0$. A key motivation for proposing this qubit encoding in [126] was that the coherent state lifetime, or T_X, or the transverse relaxation lifetime in the Kerr-cat qubit is exponentially enhanced as a function of ϵ_2, and the cost is a linear degradation of the cat state(s) lifetime, T_{YZ}. We will discuss these properties in greater detail in Chap. 3 and demonstrate it experimentally in Chap. 4.

The Kerr-cat qubit marked a paradigm shift in two concrete ways:

1. The prized longest lived mode is also the most nonlinear mode of the system, it is the weakly nonlinear bosonic oscillator. This is orthogonal to other bosonic code encodings such as the Gottesmann-Kitaev-Preskill (GKP) [19], the binomial encoding [78], or the dissipative-cat encoding [96] where the longest lived mode is also the most linear one, and any nonlinear mode is relegated to the role of ancilla.

[4] Note that in the Bloch sphere displayed here, which is adapted from [53, 66], the even and odd parity cat-states lie along the Bloch sphere Z-axis, whereas in [126], these states lie along the X-axis.

2. The Kerr-cat qubit is the first quantum bosonic code encoding in a *driven* hamiltonian nonlinear system. Over the recent years coherent, time-periodic modulation has been established as a versatile tool for realizing novel Hamiltonians. This approach, known as Floquet engineering [46], has been employed heavily in ultracold atoms to engineer new quantum materials [3, 58], topological phases [166], Floquet time-crystals [176], many-body localization (MBL) [1], and other exotic physics. Effective dynamics emerging from "Trotterizing" drives adds a new dimension of richness [58]. However, Floquet engineering in superconducting quantum circuits had been heavily ignored until the discovery of the Kerr-cat qubit.

The Kerr-cat qubit focused on the parameter $\Delta = 0$, and the treatment was strictly restricted to the ground state manifold of the SKO. It is natural to thus ask:

1. What causes the degeneracy of the qubit states at $\Delta = 0$ in Eq. (1.9)? Is this a special property of $\Delta = 0$ or are there degeneracies to be discovered also at non-zero Δ?

1.3 Dissertation Outline

Taking quantum information processing with continuous variable systems as our motivation, we will begin straight away and introduce the effective Hamiltonian corresponding to the Kerr nonlinear oscillator submitted to a squeezing interaction in Chap. 2, which we will call the squeeze-driven Kerr oscillator (SKO). Then, we will present the various representations and properties of the SKO in Chap. 3, specifically the operator space, quantum, semi-classical, and classical phase space representations. Different representations are convenient to understand different experimental observations, such as the spectroscopy, tunnel-driven Rabi, and coherent- and cat-state lifetime measurements presented later. Next, in Chap. 4, we will introduce the experimental setup that realizes the desired SKO effective Hamiltonian in a SNAIL transmon superconducting circuit submitted to a fast-oscillating drive. We will also discuss the readout of the SKO here. The main experimental results of this thesis are presented in Chaps. 5 and 6. In Chap. 5, we present observations on the sharp nulling of tunnel-driven oscillations as a function of the frequency of the squeezing drive with a periodicity given by twice the Kerr nonlinearity. In Chap. 6, we present measurements of excited state spectroscopy of the SKO. We find the tunnel nulls in the ground state manifold are also present in the excited state manifold. Furthermore, we present measurements demonstrating an overall exponential reduction of tunneling in the excited state manifold as a function of both the drive amplitude and its frequency. These two effects contribute to the drastic enhancement of the transverse relaxation lifetime of the logical qubit encoded in the ground state manifold of our oscillator as we show in Chap. 6. We

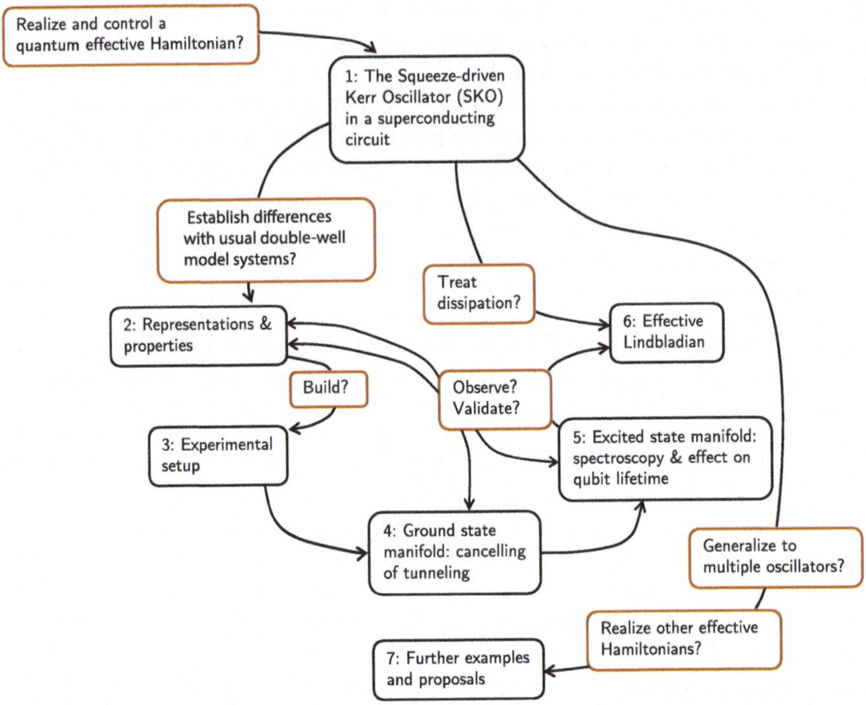

Fig. 1.4 Dissertation roadmap. Black boxes indicate main chapter titles and brown boxes indicate motivating questions. Arrows point to suggested prerequisite reading. Bidirectional arrows indicate mutual chapter dependency. The main experimental results are presented in Chaps. 5 and 6

model these lifetime measurements in Chap. 7 with an effective Lindbladian that requires going beyond the rotating wave approximation, and our treatment here is a natural extension of the effective Hamiltonian theory. We discuss future directions, present proposals and open questions in Chap. 8.

Chapter 2
The Squeeze-Driven Kerr Oscillator (SKO) Implemented in a Driven Superconducting Circuit

Abstract In this chapter, we introduce the driven superconducting tunnel junction circuit in which we experimentally engineer the squeeze-driven Kerr oscillator (SKO) effective Hamiltonian. First, we introduce the superconducting circuit in the absence of the drive. Then, we will examine its properties when submitted to a fast-oscillating drive, thus realistically modeling our experiment. The main result of this chapter is an effective Hamiltonian that is obtained by going beyond the rotating wave approximation and which governs the interesting dynamics of our driven superconducting circuit.

2.1 Introducing the SNAIL Transmon Superconducting Circuit

The superconducting circuit we use in our setup is a SNAIL transmon [50]. The word SNAIL is an acronym for Superconducting Nonlinear Asymmetric Inductive eLement. Before introducing the SNAIL transmon, we introduce the ordinary transmon.

A standard transmon [89, 140], a widely used superconducting qubit that predates the SNAIL transmon, consists of a Josephson junction shunted by a large capacitance. The transmon circuit Hamiltonian can be written as

$$\hat{H}_0 = 4E_C(\hat{N} - N_g)^2 - E_{\mathrm{J}} \cos\hat{\varphi} \tag{2.1}$$

with $\hat{N}$ and $\hat{\varphi}$ being the reduced Cooper-pair number operator and reduced phase operator across the junction, and $[\hat{\varphi}, \hat{N}] = i\hbar$, and the parameter $N_g = C_g V_g/2e$ is the gate-offset charge across the junction.

Because the Josephson potential is an even function of the superconducting phase difference φ, the nonlinearity provided in an ordinary transmon, to lowest order, is φ^4, and thus the transmon can be thought of, to lowest order, as a dipole circuit element with fourth-rank nonlinearity, which implements four-wave mixing. This four-wave mixing in turn gives rise to the Kerr effect, and ac Stark shift frequency

J. Venkatraman, *Controlling the Effective Hamiltonian of a Driven Quantum Superconducting Circuit*, Springer Theses,
https://doi.org/10.1007/978-3-031-83270-3_2

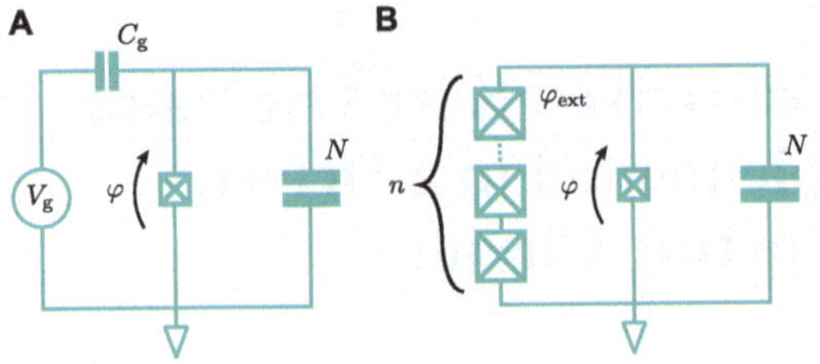

Fig. 2.1 Circuit schematics of an ordinary transmon (panel **a**) and a SNAIL transmon (panel **b**). Note that the gate voltage is omitted in the SNAIL transmon, as the shunting array of Josephson junctions provides a conducting path to null out the offset charge across the small junction

dressing of the ordinary transmon as we will see in Sect. 2.3. See Fig. 2.1a for a circuit representation of an ordinary transmon. The circuit system of the ordinary transmon finds an analogue in the mechanical system of a quantum pendulum [89, 140] and the correspondence between the systems in discussed in Appendix A. Thus, we call the term proportional to E_C the kinetic energy part and the term proportional to E_J the potential energy part of the transmon.

A SNAIL transmon [50] consists of an ordinary transmon that is further shunted by an array of Josephson junctions. The resultant loop comprising of N large Josephson junctions and a single smaller junction, with Josephson energies $E_J, \alpha E_J$ respectively, can be threaded by a dc magnetic field Φ_{ext}. At non-zero applied magnetic flux, the potential of the SNAIL transmon, to lowest order, is φ^3, and thus the SNAIL transmon can be thought of, to lowest order, as a dipole circuit element with third-rank nonlinearity, which implements three-wave mixing. Importantly, depending on the value of the dc magnetic field, the strength of the three-wave mixing and four-wave mixing can be independently tuned. This independent tunability of the Kerr nonlinearity without diminishing the three-wave mixing capabilities will prove to be critical in our experimental implementation of the SKO as we discuss in Chap. 4. For a circuit representation of the SNAIL transmon, see Fig. 2.1b. We write the SNAIL transmon Hamiltonian as

$$\hat{H}_0 = 4E_C\hat{N}^2 - \alpha E_J \cos(\hat{\varphi} - \varphi_{ext}) - nE_J \cos(\hat{\varphi}/n). \tag{2.2}$$

The circuit system of the SNAIL transmon finds an analogue in the mechanical system of two pendula suspended by rotating gears whose gear ratios is n and where the pendulum bob mass ratios is α. Thus, we call the term $\propto E_C$ the kinetic energy part and the rest the potential energy part of the SNAIL transmon. Equation (2.2) assumes a uniform phase drop across every junction in the array, and the absence of phase slips in the array junctions. This is valid when the Josephson energy is much larger than the shunting capacitive energy for all the junctions [50, 107]. See Fig. 2.2 for a sample scanning electron microscope (SEM) image of a SNAIL transmon.

Performing a Taylor expansion of the potential energy part of Eqs. (2.1) and (2.2) about the potential minimum φ_{min}, which is 0 for a standard transmon and

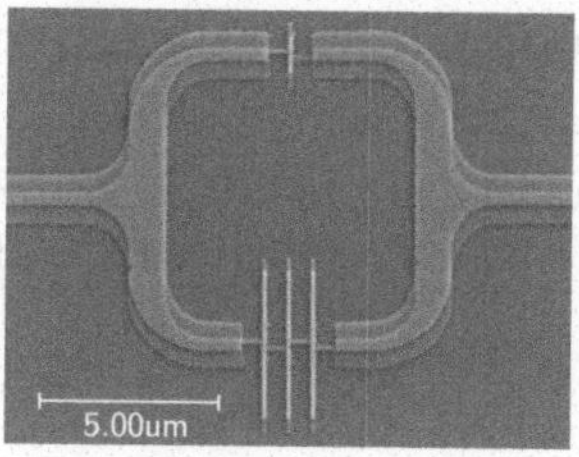

Fig. 2.2 Scanning electron microscope (SEM) image of a SNAIL superconducting circuit. For this device, $n = 3$ and $\alpha \approx 0.1$ in Eq. (2.2). Note that the device presented in this figure was not used in the experimental work presented in Chaps. 4–6. For a schematic of the device measured in the experiment, which consists of an array of two SNAIL-transmons, see Fig. 4.1

Table 2.1 Circuit parameters to bosonic oscillator parameters translation table

Parameter	Ordinary transmon	SNAIL transmon
φ_{zps}	$\left(\frac{2E_C}{E_J}\right)^{1/4}$	$\left(\frac{2E_C}{E_J(\alpha c_e + c_e'/n)}\right)^{1/4}$
ω_o	$\sqrt{8E_C E_J}$	$\sqrt{8E_C E_J\left(\alpha c_e + c_e'/n\right)}$
g_{2m}	$\frac{(-1)^{m+1}}{2(2m-1)!}\omega_o\varphi_{\text{zps}}^{2m-2}$	$\frac{(-1)^{m+1}}{2(2m-1)!}\frac{\alpha c_e + c_e'/n^{2m-2}}{\alpha c_e + c_e'/n}\omega_o\varphi_{\text{zps}}^{2m-2}$
g_{2m+1}	0	$\frac{(-1)^{m+1}}{2(2m)!}\frac{\alpha c_o + c_o'/n^{2m-1}}{\alpha c_e + c_e'/n}\omega_o\varphi_{\text{zps}}^{2m-1}$

numerically obtained for a SNAIL transmon, and further expressing $\hat{\varphi}$ and $\hat{N}$ in terms of creation and annihilation operators, $\hat{a}^\dagger$ and $\hat{a}$, allows for the Hamiltonian in Eqs. (2.1) and (2.2) to be written in the bosonic basis as,

$$\hat{H}_0 = \omega_o \hat{a}^\dagger \hat{a} + \sum_{m \geqslant 3} \frac{g_m}{m}(\hat{a} + \hat{a}^\dagger)^m,$$
$$\hat{a} = \frac{1}{2}\left(\frac{\hat{\varphi}}{\varphi_{\text{zps}}} + i\frac{\hat{N}}{N_{\text{zps}}}\right), \tag{2.3}$$

where $\varphi_{\text{zps}} N_{\text{zps}} = 1/2$, with N_{zps} denoting the zero point spread of the charge wavefunction, $c_o = \sin\varphi_{\text{min}}$, $c_e = \cos\varphi_{\text{min}}$, $c_o' = \sin((\varphi_{\text{min}} - \varphi_{\text{ext}})/M)$, and $c_e' = \cos((\varphi_{\text{min}} - \varphi_{\text{ext}})/M)$. For expressions for ω_o, g_m, and φ_{zps} from the circuit parameters of Eqs. (2.1) and (2.2), see Table 2.1.

Equation (2.3) models the general bosonic oscillator system, and forms the archetype of many Josephson nonlinear circuits, which includes the driven rf-SQUID [31], c-shunt flux qubit [171], among others.

Having gone from the circuit representation to the bosonic oscillator representation, we are now prepared to add the driving term and perform successive frame transformations to find the effective Hamiltonian.

2.2 A SNAIL Transmon Circuit Submitted to a Rapid Drive

To compute the static effective squeeze-driven Kerr oscillator Hamiltonian, we consider the Hamiltonian of a periodically driven SNAIL transmon oscillator,

$$\begin{aligned}\hat{H}(t) &= \hat{H}_0 + \hat{H}_d(t),\\ &= \omega_o \hat{a}^\dagger \hat{a} + \sum_{m\geqslant 3} \frac{g_m}{m}(\hat{a}+\hat{a}^\dagger)^m - iN_{\rm zps}\Omega_d \cos\omega_d t(\hat{a}-\hat{a}^\dagger),\end{aligned} \tag{2.4}$$

where the oscillator is characterized by $\hat{H}_0$ given by Eq. (2.3) with the SNAIL transmon coefficients provided in Table 2.1. The periodic drive is characterized by $\hat{H}_d(t)$ with the parameters Ω_d and ω_d characterizing the drive amplitude and frequency. Note that the drive coupling to the charge operator $\hat{N} = -iN_{\rm zps}(\hat{a}-\hat{a}^\dagger)$ is motivated to model the engineered capacitive coupling in the experiment. We focus on the case of a periodic drive for simplicity but we note that our treatment can be generalized to include quasiperiodic or non-monochromatic drives. Furthermore, to generate the desired squeezing interaction in the SKO, the second subharmonic of the drive at frequency $\omega_d/2$ has to lie in the vicinity of the oscillator resonance, i.e., $\omega_d/2 \approx \omega_o$.

In general, a driven nonlinear system, such as the periodically driven SNAIL transmon Eq. (2.4), does not admit closed form solutions for its time evolution. But remarkably, under a rapid drive, the dynamics can be mapped to that generated by a time-independent effective Hamiltonian. This "Kamiltonian" [59] describes a *slow* dynamics of the system, corrected only perturbatively by a *fast* micromotion. Over the last century, different perturbation methods have been developed to construct such effective Hamiltonians and have succeeded in explaining several important nonlinear dynamical phenomena [10, 44, 45, 69, 116, 174]. However, these perturbation methods can hardly be carried out beyond the lowest orders in practice and a clear understanding of the connection between many of these methods is missing [17, 130]. The differences are exacerbated by the wide disparity in starting points of the classical [15, 91, 116] and quantum methods [23, 45, 46, 111, 113, 121, 169, 174].

In Appendix B, we construct a time-independent Kamiltonian $\hat{\mathcal{K}}$ perturbatively by seeking a pertinent *canonical* transformation. The small parameter of the expansion is the ratio of the typical rate of evolution of the driven system to the frequency of the driving force. We present a recursive formula for the Kamiltonian that allows its calculation to arbitrary order and is well-suited for symbolic manipulation. It can be applied indifferently to the classical and quantum cases, the change involving only a low-level subroutine of the symbolic algorithm. Our result unifies existing methods that have been developed solely in either the classical or quantum regimes.

We illustrate our formulation by applying our recursive formula Eq. (B.9a) to the rapidly driven SNAIL circuit Hamiltonian in Sect. 2.3.

Preparatory Frame Transformations

Before performing the perturbative analysis presented in Appendix B to find the effective Hamiltonian, we perform two preparatory frame transformations on the Hamiltonian Eq. (2.4) of the SNAIL circuit Hamiltonian. These frame transformations will yield a Hamiltonian whose typical rate of evolution is much smaller than the frequency of the driving force.

Displacement Transformation

The first is a displacement unitary transformation. Specifically, the oscillator oscillates in phase with the drive, at the drive frequency. We reference the oscillator to the trivial linear response by doing $\hat{a} \to \hat{a} + \alpha_{\text{lin}}(t)$, thus bringing the concerned nonlinear dynamics of the oscillator into focus. This displacement is achieved by the unitary operator $\hat{U}(t) = \exp(\alpha_{\text{lin}}\hat{a}^\dagger - \alpha^*_{\text{lin}}\hat{a})$, where α_{lin} is a time-dependent function that is to be determined. Under the time-dependent unitary, from Eq. (B.5), we have, up to coordinate-independent terms:

$$
\begin{aligned}
&\hat{H} \to \hat{U}^\dagger(\hat{H} - i\partial_t)\hat{U} \\
&\omega_o \hat{a}^\dagger \hat{a} + \sum_{m \geqslant 3} \frac{g_m}{m}(\hat{a} + \hat{a}^\dagger)^m \\
&- iN_{\text{zps}}\Omega_d \cos\omega_d t(\hat{a} - \hat{a}^\dagger) \to \\
&\qquad \omega_o(\hat{a}^\dagger + \alpha^*_{\text{lin}})(\hat{a} + \alpha_{\text{lin}}) + \sum_{m \geqslant 3} \frac{g_m}{m}(\hat{a} + \hat{a}^\dagger + \alpha_{\text{lin}} + \alpha^*_{\text{lin}})^m \\
&\qquad - iN_{\text{zps}}\Omega_d \cos\omega_d t(\hat{a} - \hat{a}^\dagger) + i(\dot{\alpha}\hat{a}^\dagger - \alpha^*\hat{a}).
\end{aligned}
\tag{2.5a}
$$

Now, to find the linear response, we set $g_m = 0$, collect the coefficients of $\hat{a}^\dagger$, and solve for α_{lin} as

$$
\dot{\alpha}_{\text{lin}} = +i\omega_o\alpha_{\text{lin}} - N_{\text{zps}}\Omega_d \cos\omega_d t, \tag{2.5b}
$$

which yields

$$
\alpha_{\text{lin}} = \frac{i\Omega_d N_{\text{zps}}}{2(\omega_d - \omega_o)} e^{-i\omega_d t} - \frac{i\Omega_d N_{\text{zps}}}{2(\omega_d + \omega_o)} e^{i\omega_d t}, \tag{2.5c}
$$

and furthermore, we define

$$
\begin{aligned}
\alpha_{\text{lin}} + \alpha^*_{\text{lin}} &= \frac{2i\Omega_d N_{\text{zps}}\omega_d}{\omega_d^2 - \omega_o^2} e^{-i\omega_d t} - \frac{2i\Omega_d N_{\text{zps}}\omega_d}{\omega_d^2 - \omega_o^2} e^{i\omega_d t} \\
&= \Pi e^{-i\omega_d t} - \Pi^* e^{i\omega_d t},
\end{aligned}
\tag{2.5d}
$$

where in the last line of Eq. (2.5d), we use the condition $\omega_d \approx 2\omega_o$, and further defined $\Pi = i\Omega_d N_{\text{zps}}\omega_d/(\omega_d^2 - \omega_o^2) \approx 2i\Omega_d N_{\text{zps}}/(3\omega_o)$, the drive amplitude in the displaced frame.

With this, we write the rapidly drive SNAIL Hamiltonian Eq. (2.4), after the displacement unitary transformation as

$$\hat{H}(t) = \omega_o \hat{a}^\dagger \hat{a} + \sum_{m \geqslant 3} \frac{g_m}{m} (\hat{a} + \hat{a}^\dagger + \Pi e^{-i\omega_d t} + \Pi^* e^{i\omega_d t})^m. \tag{2.5e}$$

Rotating Frame Transformation

The second unitary transformation is a rotating frame transformation of the oscillator at $\omega_d/2$. We emphasize that to perform this rotating frame transformation, one needs to know apriori that the interesting dynamics occurs at the second subharmonic of the drive frequency, which corresponds to a vicinity of the oscillator's plasma frequency ω_o. As we discuss in Chap. 3, this nontrivial dynamics corresponds classically to the period doubling bifurcation that the oscillator undergoes. The associated unitary operator is $\hat{U} = \exp(-i(\omega_d/2)t\hat{a}^\dagger\hat{a})$ transforms the displaced-frame Hamiltonian as

$$\hat{H}(t) = \Delta^{\text{bare}} \hat{a}^\dagger \hat{a} + \sum_{m \geqslant 3} \frac{g_m}{m} (\hat{a} e^{-i\omega_d t/2} + \hat{a}^\dagger e^{i\omega_d t/2} + \Pi e^{-i\omega_d t} + \Pi^* e^{i\omega_d t})^m, \tag{2.6a}$$

where $\Delta_o := \omega_o - \omega_d/2$ corresponds to the detuning between the oscillator plasma frequency and the second subharmonic of the drive. As we will see next, in the effective Hamiltonian, this bare detuning will have a dressing contribution from the mixing of nonlinear terms in Eq. (2.6a).

2.3 The Effective Hamiltonian of a Rapidly Driven SNAIL Transmon Circuit Oscillator

After the preparatory steps introduced in Sects. 2.1, 2.2, and Appendix B, we will now apply the recursive formula presented in Appendix B and find the static effective Hamiltonian of a rapidly driven SNAIL circuit oscillator. Importantly, after the preparatory frame transformations, the typical rate of evolution of Eq. (2.6a), which is controlled by Δ_o and g_m/m is much smaller than the second subharmonic of the drive $\omega_d/2 \approx \omega_o$. This condition translates to demanding $\varphi_{\text{zps}} \ll 1$, which forms the small parameter of our expansion. We recall from Table 2.1 the perturbative hierarchy of our circuit nonlinearities as $g_m = \mathcal{O}(\varphi_{\text{zps}}^{m-2})$. We further take $\Delta_o = \mathcal{O}(\varphi_{\text{zps}})$.

Order 1

At order 1 in φ_{zps}, we apply the recursive formula Eq. (B.9a) or equivalently the grid Fig. B.2, to find

$$\hat{\mathcal{K}}^{(1)} = \hat{H}^{(1)} + \dot{\hat{S}}^{(1)}. \tag{2.7a}$$

By demanding $\hat{S}^{(1)}$ to be a primitive of $-\,\mathbf{osc}(\hat{H})$, i.e.

$$\hat{S}^{(1)} = -\int dt\,\mathbf{osc}(\hat{H}), \implies \frac{\hat{S}^{(1)}}{i} = \sum_{m\neq 0} \frac{H_m}{m\omega_d/2} e^{im\omega_d t/2}. \tag{2.7b}$$

We find $\hat{\mathcal{K}}^{(1)}$ to be simply the time-average of $\hat{H}$ over one period $T = 4\pi/\omega_d$:

$$\begin{aligned}\hat{\mathcal{K}}^{(1)} &= \overline{\hat{H}^{(1)}} \\ &= \Delta_o \hat{a}^\dagger \hat{a} + \epsilon_2 \hat{a}^{\dagger 2} + \epsilon_2 \hat{a}^2,\end{aligned} \tag{2.7c}$$

where we recall $\Delta_o = \omega_o - \omega_d/2$, $\epsilon_2 = g_3\Pi$. In Eq. (2.7c), we have taken ϵ_2 to be real without loss of generality: the absolute phase of the drive is an arbitrary choice and determines the phase of ϵ_2. Note that taking the ordinary rotating wave approximation (RWA) over Eq. (2.4) will yield the order 1 Kamiltonian Eq. (2.7c). In this sense the RWA corresponds to the lowest order result of our recursive formula Eq. (B.9a), which when carried out to all orders corresponds to a canonical transformation. Note that, for simplicity, we have set the integration constant of $S^{(1)}$ to be zero and so we shall do for higher orders.

It is important to remark that the operators $\hat{a}^\dagger \hat{a}$, $\hat{a}^{\dagger 2}$, and $\hat{a}^2$ form a Lie algebra, which gives rise to important mathematical applications [81].

Order 2

Continuing the calculation, at order 2 in φ_{zps}, using the grid Fig. B.2, we have:

$$\hat{\mathcal{K}}^{(2)} = \hat{H}^{(2)} + \dot{\hat{S}}^{(2)} + \frac{1}{i}[\hat{S}^{(1)}, \hat{H}^{(1)}] + \frac{1}{2!}\frac{1}{i}[\hat{S}^{(1)}, \dot{\hat{S}}^{(1)}]. \tag{2.8a}$$

By plugging in the expression of $\hat{S}^{(1)}$ found from Eq. (2.7b) into Eq. (2.8a) and demanding that $\hat{S}^{(2)} = -\int dt\,\mathbf{osc}(\hat{H}^{(2)} + \frac{1}{i}[\hat{S}^{(1)}, \hat{H}^{(1)}] + \frac{1}{2!i}[\hat{S}^{(1)}, \dot{\hat{S}}^{(1)}])$ we find the following time-independent expression for $\hat{K}^{(2)}$ as:

$$\begin{aligned}\hat{\mathcal{K}}^{(2)} &= \Delta^{(2)} \hat{a}^\dagger \hat{a} - K^{(2)} \hat{a}^{\dagger 2} \hat{a}^2 \\ &= (\Delta^{(2)}_{\text{Lamb}} + \Delta^{(2)}_{\text{ac}}) \hat{a}^\dagger \hat{a} - K^{(2)} \hat{a}^{\dagger 2} \hat{a}^2,\end{aligned} \tag{2.8b}$$

where

$$
\begin{aligned}
\Delta^{(2)}_{\text{Lamb}} &= 3g_4 - \frac{20g_3^2}{3\omega_o} \\
\Delta^{(2)}_{\text{ac}} &= \left(6g_4 - \frac{9g_3^2}{\omega_o}\right)|\Pi|^2 \\
K^{(2)} &= \frac{10g_3^2}{3\omega_o} - \frac{3g_4}{2}.
\end{aligned} \tag{2.8c}
$$

In Eq. (2.8c), $\Delta^{(2)}_{\text{Lamb}}$ corresponds to the *Lamb shift* dressing at order 2, i.e., dressing of the plasma frequency of the oscillator due to the non-commutativity of $\hat{a}$ and $\hat{a}^\dagger$. To distinguish the Lamb-shifted frequency of the oscillator from the plasma frequency ω_o, we introduce $\omega_a = \omega_o + \Delta_{\text{Lamb}}$, and this corresponds to the measured transition frequency between the lowest eigenstates of the SNAIL circuit oscillator. The parameter Δ_{ac} is the *ac Stark shift* frequency dressing, i.e., the drive-power dependent frequency shift of the oscillator. At order 2, this term is linear in the drive power but acquires nonlinear corrections at higher orders in perturbation theory as we will see next. The parameter K corresponds to the *Kerr nonlinearity* of the oscillator. We have taken the sign to be negative because in transmon-like superconducting circuits $g_3 = 0$, and $g_4 < 0$, thus by the chosen convention K conveniently takes a positive value. However as is evident in Eq. (2.8c), for the SNAIL transmon, the relative signs of g_4 and g_3^2/ω_o determines the sign of K.

Equation (2.8c) already highlights why it is critical to go beyond the RWA to find the effective Hamiltonian of the rapidly driven SNAIL transmon, and why simply expanding the Josphson junction potential to a high Taylor series order is insufficient. By only Taylor expanding the potential and performing the RWA, the term $\frac{10g_3^2}{3\omega_o}$ contributing towards the Kerr nonlinear coefficient would be missed out, and an attempt of extracting the Kerr nonlinear coefficient of the SNAIL transmon would yield incorrect results [144]. In this sense, to capture the correct coefficients in the effective Hamiltonian, one needs to both Taylor expand the potential to high enough order and account for the beyond RWA terms from lower orders [153, 170]. Apart from the method presented in Appendix B, we have developed a diagrammatic approach to compute the effective Hamiltonian of drive Josephson oscillators. A discussion on this is beyond the scope of this dissertation, but will be exhaustively covered in [170].

Order 3

Applying the recursive formula Eq. (B.9a) at order 3, we find

$$
\begin{aligned}
\hat{\mathcal{K}}^{(3)} &= \Delta^{(3)}\hat{a}^\dagger\hat{a} - K^{(3)}\hat{a}^{\dagger 2}\hat{a}^2 + \hat{a}^{\dagger 2}\hat{\epsilon}_2^{(3)} + \hat{\epsilon}_2^{\dagger(3)}\hat{a}^2 \\
&= \Delta^{(3)}\hat{a}^\dagger\hat{a} - K^{(3)}\hat{a}^{\dagger 2}\hat{a}^2 + \hat{a}^{\dagger 2}(\epsilon^{(3)}_{2,\Pi} + \epsilon^{(3)}_{2,\hat{n}}\hat{a}^\dagger\hat{a}) + (\epsilon^{*(3)}_{2,\Pi}\hat{a}^2 + \epsilon^{*(3)}_{2,\hat{n}}\hat{a}^\dagger\hat{a})\hat{a}^2
\end{aligned} \tag{2.9a}
$$

with $\Delta^{(3)} = \Delta^{(3)}_{\text{Lamb}} + \Delta^{(3)}_{\text{ac}}$, where

$$\Delta^{(3)}_{\text{Lamb}} = \frac{20}{3}\frac{\Delta_o g_3^2}{\omega_o^2}$$

$$\Delta^{(3)}_{\text{ac}} = \frac{17}{2}\frac{\Delta_o g_3^2}{\omega_o^2}|\Pi|^2$$

$$K^{(3)} = -\frac{10}{3}\frac{\Delta_o g_3^2}{\omega_o^2}$$

$$\epsilon^{(3)}_{2,\Pi} = \left(6g_5 - \frac{141}{10}\frac{g_3 g_4}{\omega_o} + \frac{221}{180}\frac{g_3^3}{\omega_o^2}\right)|\Pi|^2\Pi + \left(6g_5 - \frac{63}{4}\frac{g_3 g_4}{\omega_o} + \frac{1}{3}\frac{g_3^3}{\omega_o^2}\right)\Pi$$

$$\epsilon^{(3)}_{2,\hat{n}} = \left(4g_5 - \frac{21}{2}\frac{g_3 g_4}{\omega_o} + \frac{2}{9}\frac{g_3^3}{\omega_o^2}\right)\Pi. \tag{2.9b}$$

In Eq. (2.9a), we have introduced an operator-valued squeezing-drive amplitude $\hat{\epsilon}_2^{(2)}$. It includes a photon number-independent part and a photon-number dependent part, i.e., $\hat{\epsilon}_2^{(2)} = \epsilon^{(2)}_{2,\Pi} + \epsilon^{(2)}_{2,\hat{n}}\hat{a}^\dagger\hat{a}$. With this notation, it is clear that the effective Hamiltonian up to the third order contains only three types of terms: harmonic-type proportional to $\hat{a}^\dagger\hat{a}$, Kerr-type proportional to $\hat{a}^{\dagger 2}\hat{a}^2$, or squeezing-type proportional to $(\hat{a}^{\dagger 2} + \hat{a}^2)$.[1] It is worth stressing on this point because the main challenge with engineering parametric interactions by driving a superconducting circuit is that unwanted interactions accompany and often proliferate uncontrollably along with the desired interaction as soon as one goes beyond the RWA [138, 170]. But remarkably, in the SKO, these uncontrolled interactions are absent even beyond the RWA. The first occurrence of a term that is neither a squeezing, nor a harmonic, nor a Kerr-type term, will appear at order 4 as we see next.

Order 4

At order 4, we find

$$\begin{aligned}\hat{\mathcal{K}}^{(4)} &= \Delta^{(4)}\hat{a}^\dagger\hat{a} - K^{(4)}\hat{a}^{\dagger 2}\hat{a}^2 + K_6^{(4)}\hat{a}^{\dagger 3}\hat{a}^3 \\ &\quad + \hat{a}^{\dagger 2}(\epsilon^{(4)}_{2,\Pi} + \epsilon^{(4)}_{2,\hat{n}}\hat{a}^\dagger\hat{a}) + (\epsilon^{*(4)}_{2,\Pi} + \epsilon^{*(4)}_{2,\hat{n}}\hat{a}^\dagger\hat{a})\hat{a}^2 + \epsilon_4^{(4)}\hat{a}^{\dagger 4} + \epsilon_4^{*(4)}\hat{a}^4,\end{aligned} \tag{2.10a}$$

with $\Delta^{(4)} = \Delta^{(4)}_{\text{Lamb}} + \Delta^{(4)}_{\text{ac}}$, where

[1] Note that the term proportional to $(\hat{a}^{\dagger 2} + \hat{a}^2)$ is also referred to as a pairing interaction/two-photon drive in the condensed matter physics/nuclear physics communities [79] in contrast with $(\hat{a}^\dagger\hat{a})^2$, which is referred to as the photon-photon interaction.

$$\Delta^{(4)}_{\text{Lamb}} = 15g_6 - \frac{220}{3}\frac{g_3g_5}{\omega_o} - 18\frac{g_4^2}{\omega_o} + 188\frac{g_3^2g_4}{\omega_o^2} - \frac{20}{3}\frac{\Delta_o^2 g_3^2}{\omega_o^3} \tag{2.10b}$$

$$\Delta^{(4)}_{\text{ac}} = \left(30g_6 - \frac{644}{5}\frac{g_3g_5}{\omega_o} + 9\frac{g_4^2}{\omega_o} + \frac{15113}{150}\frac{g_3^2g_4}{\omega_o^2} - \frac{2821}{540}\frac{g_3^4}{\omega_o^3}\right)|\Pi|^4 \tag{2.10c}$$

$$+\left(60g_6 - 232\frac{g_3g_5}{\omega_o} - \frac{108}{5}\frac{g_4^2}{\omega_o} + \frac{1342}{5}\frac{g_3^2g_4}{\omega_o^2}\right.$$

$$\left. + \frac{982}{45}\frac{g_3^4}{\omega_o^3} - \frac{33}{4}\frac{\Delta_o^2 g_3^2}{\omega_o^3}\right)|\Pi|^2 \tag{2.10d}$$

$$K^{(4)} = (-30g_6 + 116\frac{g_3g_5}{\omega_o} + \frac{54}{5}\frac{g_4^2}{\omega_o} - \frac{671}{5}\frac{g_3^2g_4}{\omega_o^2} + \frac{113}{90}\frac{g_3^4}{\omega_o^3})|\Pi|^2 \tag{2.10e}$$

$$+(-15g_6 + 84\frac{g_3g_5}{\omega_o} + \frac{153}{8}\frac{g_4^2}{\omega_o} - 225\frac{g_3^2g_4}{\omega_o^2} + \frac{10}{3}\frac{\Delta_o^2g_3^2}{\omega_o^3} - \frac{1570}{9}\frac{g_3^4}{\omega_o^3}) \tag{2.10f}$$

$$K_6^{(4)} = \frac{10}{3}g_6 - \frac{56}{3}\frac{g_3g_5}{\omega_o} - \frac{17}{4}\frac{g_4^2}{\omega_o} + 50\frac{g_3^2g_4}{\omega_o^2} \tag{2.10g}$$

$$\epsilon^{(4)}_{2,\Pi} = \left(\frac{10669}{600}\frac{\Delta_o g_3g_4}{\omega_o^2} + \frac{46313}{5400}\frac{\Delta_o g_3^3}{\omega_o^3}\right)|\Pi|^2\Pi$$

$$+\left(\frac{353}{16}\frac{\Delta_o g_3g_4}{\omega_o^2} + \frac{521}{36}\frac{\Delta_o g_3^3}{\omega_o^3}\right)\Pi \tag{2.10h}$$

$$\epsilon^{(4)}_{2,\hat{n}} = \left(\frac{353}{24}\frac{\Delta_o g_3g_4}{\omega_o^2} + \frac{521}{54}\frac{\Delta_o g_3^3}{\omega_o^3}\right)\Pi \tag{2.10i}$$

$$\epsilon^{(4)}_4 = \frac{5}{2}g_6 + \frac{2}{15}\frac{g_3g_5}{\omega_o} - \frac{33}{4}\frac{g_4^2}{\omega_o} - \frac{101}{24}\frac{g_3^2g_4}{\omega_o^2} + \frac{2009}{162}\frac{g_3^4}{\omega_o^3}. \tag{2.10j}$$

At order 4, we see the first terms proportional to $(\hat{a}^{\dagger 4} + \hat{a}^4)$. Indeed, these would occur for strong drives and are undesirable terms in the SKO effective Hamiltonian Eq. (1.9). Yet the coefficient of this term has a nontrivial dependence on the circuit oscillator's bare nonlinearities. By smartly engineering the bare nonlinearities of the circuit or introducing more knobs other than the external flux to tune them, one may null the effect of these terms. This level of circuit engineering is beyond the scope of the current work. With the recursive formula Eq. (B.9a) and the grid Fig. B.2, one has the tools to make informed decisions for circuit design. We used a computer algebra system [146] to automate the calculation displayed above. Although we will stop the calculation here, one can in principle continue to higher orders. Our formulation [153] overcomes limitations of existing time-dependent perturbation methods [130] allowing computations beyond the RWA that were impossible before.

In conclusion, we have theoretically examined in detail, how rapidly driving a SNAIL transmon superconducting circuit yields an effective Hamiltonian of the squeeze-driven Kerr oscillator (SKO) Eq. (1.9), the main object of interest. Treating the driven circuit beyond the RWA, we saw the emergence of terms which simply dressed the coefficients Δ, K, and ϵ_2 in Eq. (1.9). By going to even higher order, we saw terms proportional to $\hat{a}^{\dagger 3}\hat{a}^3$, and $(\hat{a}^{\dagger 4} + \hat{a}^4)$. We expect the contribution from these terms to be negligible for weak drives since g_6 is proportional to $\omega_o \varphi_{\text{zps}}^4 \ll 1$, yet for sufficiently strong drives the effects of these terms may be important to consider, but this is beyond the scope of the current work. In a different work [170], we provide a comprehensive characterization of beyond-RWA phenomena, specifically multiphoton resonances, which are responsible for anamolous state transitions and play a major role in qubit readout limiting fidelities [10, 138, 142, 177].

Before discussing how we experimentally engineer and control the effective Hamiltonian Eq. (1.9), we next discuss some of its unique properties and representations. These properties will be important to understand the experimental results presented in Chap. 4.

Chapter 3
Representations and Properties of the SKO

Abstract In this chapter, we will discuss in detail the classical, semi-classical, and quantum representations of the effective Hamiltonian of the squeeze-driven Kerr oscillator (SKO), obtained by rapidly driving the SNAIL transmon superconducting circuit (see detailed modeling in Chap. 2). As we will see, different representations are convenient to understand different properties. There are two properties we will investigate as a function of the drive frequency and the drive amplitude: spectral degeneracies, and the inter-well tunneling amplitude. As we uncover the SKO's tunneling properties, we will also contrast them with the tunneling properties of an ordinary double-well potential. Relevant literature is surveyed along the way.

Let us reintroduce the effective Hamiltonian of the squeezed-driven Kerr oscillator Eq. (1.9), the main object of our interest as

$$\hat{H} = \Delta \hat{a}^{\dagger}\hat{a} - K\hat{a}^{\dagger 2}\hat{a}^{2} + \epsilon_{2}(\hat{a}^{\dagger 2} + \hat{a}^{2}), \tag{3.1}$$

where without loss of generality (WLOG) we have taken ϵ_2 to be real-valued.

3.1 Phase Space Representations of the SKO Effective Hamiltonian

We obtain the phase space formulation of Eq. (3.1) by taking the invertible Wigner transform [35] $\mathfrak{W}$ as

J. Venkatraman, *Controlling the Effective Hamiltonian of a Driven Quantum Superconducting Circuit*, Springer Theses,
https://doi.org/10.1007/978-3-031-83270-3_3

$$\begin{aligned}
&\hat{x} \rightarrow \mathfrak{W}\{\hat{x}\} = x; \qquad \hat{p} \rightarrow \mathfrak{W}\{\hat{p}\} = p;\\
&\hat{a} \rightarrow \mathfrak{W}\{\hat{a}\} = a = (x + ip)/\sqrt{2\lambda}; \quad \hat{a}^\dagger \rightarrow \mathfrak{W}\{\hat{a}^\dagger\} = a^*;\\
&\hat{a}^\dagger\hat{a} \rightarrow a^* \star a = a^*a - \frac{1}{2} = \frac{x^2 + p^2}{2\lambda} - \frac{1}{2};\\
&\hat{a}^{\dagger 2}\hat{a}^2 \rightarrow a^{*2} \star a^2 = a^{*2}a^2 - 2a^*a + \frac{1}{2}\\
&\qquad\qquad = \frac{(x^2 + p^2)^2}{4\lambda^2} - \frac{(x^2 + p^2)}{\lambda} + \frac{1}{2};\\
&\hat{a}^{\dagger 2} + \hat{a}^2 \rightarrow a^{*2} + a^2 = \frac{(x^2 - p^2)}{\lambda},
\end{aligned} \tag{3.2}$$

where the Groenewold star product [67] is given by

$$\mathfrak{W}\{\hat{A}\hat{B}\} = A \star B = A \exp\left(\frac{1}{2}(\overleftarrow{\partial}_a \overrightarrow{\partial}_{a^*} - \overleftarrow{\partial}_{a^*} \overrightarrow{\partial}_a)\right) B. \tag{3.3}$$

At this point, λ is a dimensionless rescaling parameter. We will connect it with the Hamiltonian parameters later, while discussing the classical limit ($\lambda \rightarrow 0$) of our system, and thereby give it physical significance. The Moyal bracket [35, 115] over a and a^* is defined as $\{A, B\}_{a,a^*} = A \star B - B \star A$ so that we have $\{a, a^*\} = 1$. For a pedagogical exposition on the phase space formulation of quantum mechanics, we refer the reader to Appendix E and [24, 35, 76]. With Eq. (3.2), we write Eq. (3.1) in the phase space formulation of quantum mechanics, up-to coordinate-independent terms, as

$$\begin{aligned}
H &= (\Delta + 2K)\left(\frac{x^2 + p^2}{2\lambda}\right) - K\left(\frac{x^2 + p^2}{2\lambda}\right)^2 + \epsilon_2\left(\frac{x^2 - p^2}{\lambda}\right)\\
&= \Delta_o\left(\frac{x^2 + p^2}{2\lambda}\right) - K\left(\frac{x^2 + p^2}{2\lambda}\right)^2 + \epsilon_2\left(\frac{x^2 - p^2}{\lambda}\right),
\end{aligned} \tag{3.4a}$$

where we recall that $\Delta_o = \Delta + 2K = \omega_o - \omega_d/2$ is the detuning between the oscillator plasma frequency and half the drive frequency. We further rescale Eq. (3.4a) by $-K/\lambda^2$ so as to have a coefficient of order 1 for the nonlinear term and rearrange Eq. (3.4a) as

$$\begin{aligned}
\frac{-H\lambda^2}{K} &= \left(\frac{x^2 + p^2}{2}\right)^2 - \frac{2\epsilon_2\lambda}{K}\frac{x^2}{2}\left(1 + \frac{(\Delta + 2K)}{2\epsilon_2}\right)\\
&\quad + \frac{2\epsilon_2\lambda}{K}\frac{p^2}{2}\left(1 - \frac{(\Delta + 2K)}{2\epsilon_2}\right).
\end{aligned} \tag{3.4b}$$

By choosing the scale of phase space $\lambda = K/2\epsilon_2$ Eq. (3.4b) becomes

$$\frac{-H\lambda^2}{K} = \left(\frac{x^2+p^2}{2}\right)^2 - \frac{x^2}{2}\left(1+\frac{\Delta}{2\epsilon_2}+2\lambda\right) + \frac{p^2}{2}\left(1-\frac{\Delta}{2\epsilon_2}-2\lambda\right). \tag{3.4c}$$

3.2 Classical Limit

The term proportional to λ in Eq. (3.4c) involves a commutator, and corresponds to the Lamb shift. The classical limit then consist in dropping this term. This is valid for $\lambda \ll \min(\Delta/2\epsilon_2,\ 1)$ and translates to $\Delta/K,\ \epsilon_2/K \gg 1$. The classical limit equivalently corresponds to equating $\Delta = \omega_a - \omega_d/2$ to $\Delta_o = \omega_o - \omega_d/2$, thus neglecting the Lamb shift correction to the oscillator frequency. We thus write the classical limit[1] of Eq. (3.4c) as

$$H_{\text{cl}} = -\frac{K}{\lambda^2}\left[\left(\frac{x^2+p^2}{2}\right)^2 - \frac{x^2}{2}\left(1+\frac{\Delta}{2\epsilon_2}\right) + \frac{p^2}{2}\left(1-\frac{\Delta}{2\epsilon_2}\right)\right]. \tag{3.4d}$$

For $\lambda = 1$, we get

$$\frac{H_{\text{cl}}}{K} = -\left(\frac{x^2+p^2}{2}\right)^2 + \frac{x^2}{2}\left(\frac{\Delta}{K}-\frac{2\epsilon_2}{K}\right) - \frac{p^2}{2}\left(-\frac{\Delta}{K}+\frac{2\epsilon_2}{K}\right) \tag{3.4e}$$

$$= \frac{\Delta}{K}a^*a - a^{*2}a^2 + \frac{\epsilon_2}{K}(a^{*2}+a^2), \tag{3.4f}$$

We call the surface for H in Eq. (3.4a) the *metapotential* of the squeeze-driven Kerr oscillator (SKO), and the surface obtained by taking the classical limit for H, H_{cl} in Eq. (3.4e), as the *classical metapotential.*

The classical metapotential corresponding to Eq. (3.4e) is spanned by two parameters Δ/K and ϵ_2/K. The parameter space of Δ/K and ϵ_2/K is divided by two phase transitions located at $\Delta = \pm 2\epsilon_2$. Different phases are characterized by the number of nodes in the classical metapotential and we refer to them as the single-, double-, and triple-node phases. These phases correspond to different metapotential topologies. We show them as contour line insets in Fig. 3.1a, representing classical orbits. The single-node phase occurs for $\Delta < -2\epsilon_2$, and presents only one metapotential well. For $\Delta \geq -2\epsilon_2$, the oscillator has bifurcated and the metapotential acquires two wells. In the presence of dissipation, these wells house stable nodes

[1] In the absence of dissipation, the metapotential acquires two wells as soon as $\epsilon_2,\ \Delta > 0$, i.e. there is no threshold for bifurcation of the driven oscillator. In the presence of dissipation, this threshold is finite and set by $\epsilon_2/K > \gamma/(4K)$ as derived in Eq. (3.6d).

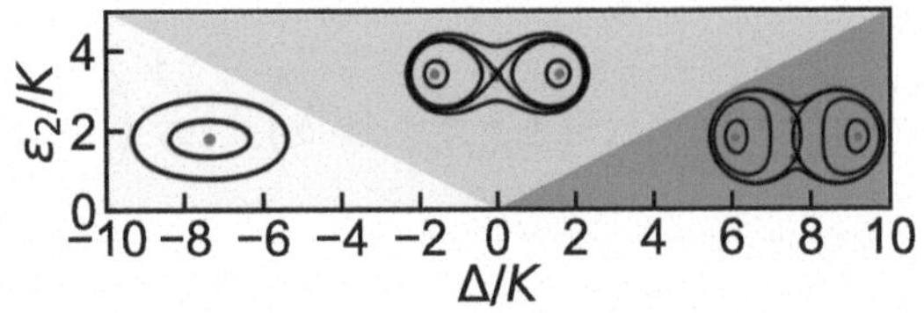

Fig. 3.1 Classical period doubling phase diagram in the parameter space of Δ/K and ϵ_2/K. Note that the classical limit corresponds to taking $\Delta = \Delta_o = \omega_o - \omega_d/2$, thus neglecting the Lamb shift correction to the oscillator frequency. Different phases marked by color are characterized by the number of nodes in the classical metapotential and we refer to them as the single-, double-, and triple-node phases. Each phase is associated with characteristic metapotential surfaces, the constant energy contour lines of which are shown as insets. The nodes and saddle points are marked in green and red respectively in this figure. The phases containing insets with more than one node correspond to regions in parameter space where the oscillator has bifurcated

(attractors). The emergent ground state manifold has been exploited, for $\Delta = 0$, in the Kerr-cat qubit [53, 66]. In the interval $-2\epsilon_2 \leq \Delta < 2\epsilon_2$, an unstable node (saddle point) appears at the origin. For $\Delta \geq 2\epsilon_2$, the saddle point at the origin splits into two saddle points and an attractor reappears at the origin. The barrier height of the classical metapotential is given by $(\Delta+2\epsilon_2)^2/4K$ in the double-node phase and by $2\epsilon_2\Delta/K$ phase.

Classical Limit in the Presence of Dissipation

In Fig. 3.2, we present the modification of the period doubling phase diagram under dissipation. In the absence of dissipation, motion on the classical metapotential surface would simply form closed contours as indicated with the insets in Fig. 3.1. In the presence of dissipation, the decay towards the origin of phase space is captured to leading order by making the Hamiltonian complex-valued as:

$$H_{\text{cl}} = \left(\Delta - i\frac{\gamma}{2}\right) a^* a - K a^{*2} a^2 + \epsilon_2 (a^{*2} + a^2), \tag{3.5}$$

and γ represents the damping rate. We sketch the procedure to obtain Eq. (3.5) from first principles here but leave the detailed derivation as an exercise. Consider a nonlinear oscillator and write down its equations of motion with Eq. (3.4f) being the Hamiltonian and an additional velocity-dependent damping term with coefficient γ. Using complex-coordinates (a, a^*), symmetrizing the dissipation, and further performing a rotating wave approximation (RWA) yields Eq. (3.5). This treatment also holds to capture the "quantum trajectory" Lindblad dynamics in the absence of quantum jumps. For a more elaborate treatment of the dissipative evolution of our experimental system, see Chap. 7. Similar classical period doubling phase diagrams can be found in [144, 167, 178].

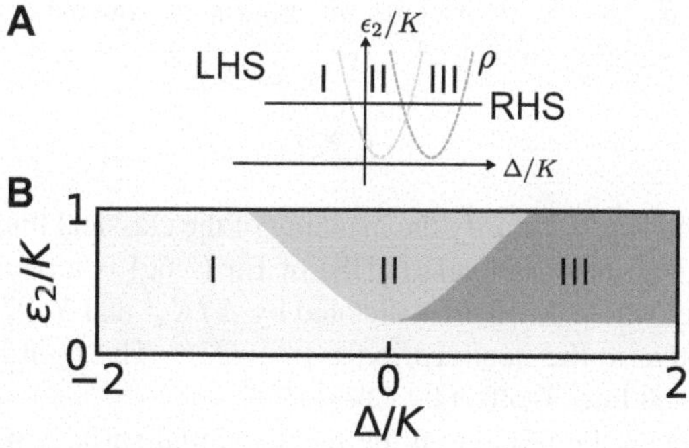

Fig. 3.2 (**a**) Schematic representation of Eq. (3.6c) in the Δ/K, and ϵ_2/K parameter space for single-node, double-node, and triple-node phases. The left-hand-side of Eq. (3.6c) represents a parabola in ρ and the right-hand-side of Eq. (3.6c) represents a constant line. (**b**) Classical period doubling phase diagram in the parameter space of Δ/K and ϵ_2/K for $\gamma/K = 1$. The phase marked I represents the single-node phase where the associated classical metapotential has a single extremum. The phase marked II represents the double-node phase where the oscillator has bifurcated and the classical metapotential has two nodes, with the origin being a saddle point. The phase marked III represents the triple-node phase where the associated classical metapotential has three nodes and two saddle points, with the origin being a node. The nodes and saddle points are marked in green and red respectively in Fig. 3.1

To find the location of the attractors in the presence of dissipation, we find the derivative of the complex-Hamiltonian in steady state and set it to zero as

$$\begin{aligned} i\dot{a} &= \partial_{a^*} H_{\rm cl} = 0 \\ &\implies \left(\Delta - i\frac{\gamma}{2}\right) a - 2Ka^* a^2 + 2\epsilon_2 a^* = 0. \end{aligned} \tag{3.6a}$$

Defining $a = re^{i\theta}$, and $\rho = r^2$, we solve for r as

$$\left(\Delta - i\frac{\gamma}{2}\right) r + 2\epsilon_2 r e^{-2i\theta} = 2Kr^3. \tag{3.6b}$$

The solution $r = 0$ corresponding to the origin is one solution for Eq. (3.6b). The other nontrivial solution is given by solving

$$\left(\frac{\Delta}{K} - 2\rho\right)^2 + \left(\frac{\gamma/2}{K}\right)^2 = \left(\frac{2|\epsilon_2|}{K}\right)^2, \tag{3.6c}$$

The nontrivial solutions for ρ is given as

$$\rho = \frac{\Delta}{K} \mp \sqrt{\left(\frac{2\epsilon_2}{K}\right)^2 - \left(\frac{\gamma}{2K}\right)^2} \tag{3.6d}$$

Nontrivial solutions to ρ will exist iff the nonlinear squeezing drive amplitude counteracts the dissipation strength, i.e.,

$$4|\epsilon_2| \geq \gamma. \tag{3.6e}$$

Equation (3.6c) and Fig. 3.2 clarify the meaning of the classical limit in the presence of dissipation γ. The left hand side (LHS) of Eq. (3.6c) is a parabola in ρ in the $\Delta/K, \epsilon_2/K$ plane, whose X-offset is dictated by Δ/K,[2] and Y-offset is dictated by the damping relative to the nonlinearity: $((\gamma/2)/K)^2$. The right hand side (RHS) represents a constant line, Y-offset by $(2|\epsilon_2|/K)^2$.

To formally obtain the motion on the metapotential surface, it is convenient to decompose $H_{\rm cl}$ into its constituent real and imaginary parts, i.e., $H_{\rm cl} = \Re(H_{\rm cl}) + \Im(H_{\rm cl})$. The velocity vector $\vec{v}$ of any point (x, p) on such a surface is thus given by

$$\begin{aligned} \vec{v} = (\dot{x}, \dot{p}) &= \left(\frac{\partial \Re(H_{\rm cl})}{\partial p}, -\frac{\partial \Re(H_{\rm cl})}{\partial x}\right) + \left(\frac{\partial \Im(H_{\rm cl})}{\partial p}, \frac{\partial \Im(H_{\rm cl})}{\partial p}\right) \\ &= \nabla \Re(H_{\rm cl}) \times \hat{n} + \nabla \Im(H_{\rm cl}) \end{aligned} \tag{3.7}$$

where $\hat{n}$ represents a unit vector normal to the (x, p) plane and x, p, and $\hat{n}$ form a right-handed three-dimensional coordinate system.

In Fig. 3.3, we draw the velocity field under Eq. (3.5) for the period-doubling bifurcation, in the triple-node phase for $\Delta/K = 11, \gamma/K = 0.1$, and $\epsilon_2/K = 4$. The coordinates $x = (a + a^*)/\sqrt{2}$ and $p = -i(a - a^*)/\sqrt{2}$ represent the position and momentum in a time-averaged displaced frame rotating at the second subharmonic of the drive, $\omega_d/2$. The color represents the magnitude of the velocity vector and arrows represent flow lines. The velocity vector $\vec{v} = (\dot{x}, \dot{p})$ of any point within the metapotential $H_{\rm cl}$ can be broken down as the sum of two velocity components as shown in Eq. (3.7). In a, we show the component of the velocity field that is attributed to the real part of $H_{\rm cl}$: $\nabla\Re(H_{\rm cl}) \times \hat{n}$ where $\hat{n}$ represents a unit vector perpendicular to the direction of flow. This component, which informs about the non-dissipative flow, yields stationary orbits around the nodes and saddles, marked in green and red respectively. (b) Component of the velocity field that is attributed to the imaginary part of $H_{\rm cl}$: $\nabla\Im(H_{\rm cl})$. To leading order here, dissipation counteracts the bifurcation, causing the velocity vectors to point radially inward; however it is possible to create nontrivial bifurcated steady states, not only in $\Re(H_{\rm cl})$ but also in $\Im(H_{\rm cl})$ by dissipation engineering [96, 114]. In Fig. 3.3c, we show the total velocity field obtained by vector addition of Fig. 3.3a and b. When dissipation is accounted for, the node transforms to an attractor, since neighboring points in phase space are drawn towards it. Moreover, the vectorial addition dresses the points of the nodes

[2] The only change in this analysis when the ac Stark shift into account is that Δ undergoes a dressing.

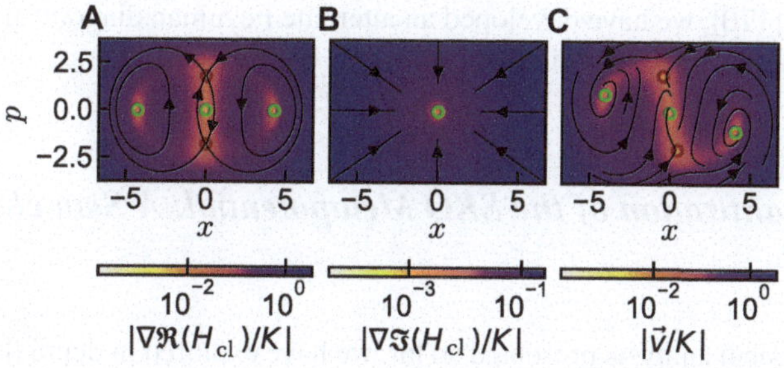

Fig. 3.3 (**a**) Real, (**b**) imaginary components, and (**c**) total velocity field representing the full nonlinear dynamics on the classical metapotential surface Eq. (3.7) of the SKO. The parameters are taken to be $\Delta/K = 11, \gamma/K = 0.1$, and $\epsilon_2/K = 4$. From (**a**)–(**c**), the color corresponds to the magnitude of the velocity vector and the arrows show its flow lines. The coordinates x and p correspond to the position and momentum in a time-averaged displaced frame rotating at half the drive frequency. (**a**) Flow lines of the velocity field attributed to the real part of the classical metapotential. In the absence of dissipation, the flow lines reduce to stationary orbits, with the arrow representing the direction of flow along the orbit. The positions of the nontrivial nodes and saddles are marked in green and red respectively. (**b**) Flow lines of the velocity component attributed to the imaginary part of the metapotential. Under the driving modeled Chap. 2, to leading order, the bifurcation only occurs in the real part of the metapotential, and thus the imaginary part only has an extremum at $x = 0$, $p = 0$, however, there are driven-dissipative operations that create bifurcations in this component as well [96]. (**c**) Vectorial sum of (**a**) and (**b**) according to Eq. (3.7) that puts the real and imaginary parts of the metapotential on equal footing. Dissipation causes the nodes (saddle) to attract (repel) neighboring trajectories in phase space, and to leading order only causes them to acquire a phase given by $\theta = \frac{1}{2} \arctan -\gamma/(2(\Delta - 2K\rho))$

and saddles so that the nodes on the X-axis in Fig. 3.3a , now acquire a phase given as $\tan 2\theta = -\gamma/2(\Delta - 2K\rho)$.

To summarize, the metapotential governs the effective dynamics of the driven-(dissipative)-nonlinear quantum system. Furthermore, as customary, we plot $-H$ rather than H to respect the familiar notion that in the presence of dissipation, stable equilibria correspond to well-bottoms rather than hill-tops.

The name metapotential warrants a special name because unlike a true potential whose gradient yields a force, the gradient of the metapotential yields a velocity as we discuss next. Hence, counter-intuitively, both its maximum and minimum extremal points could attract neighboring points in phase space unlike a true potential function. Although this object has been used in previous nonlinear dynamics literature [44, 116, 167, 168], to our best knowledge, tools to compute the metapotential and its classical limit have been developed for the simplest nonlinear dynamics phenomena to the lowest orders such as the period doubling bifurcation, but these methods are too tedious to be generalized. The value of the treatment in Appendix B is its ease of extension to higher order bifurcations that occur in a driven-dissipative nonlinear oscillator. This object not only reveals intuition about these novel dynamical phenomena, but computing the metapotential is also of practical value; constructing the metapotential forms the first step to be accomplished before actively controlling/engineering it in experimental systems

[53, 154]. In [170], we have developed an alternate Feynman diagram-like approach to compute the effective Hamiltonian of driven bosonic oscillators.

Action Quantization of the SKO Metapotential: A Semiclassical Treatment

With the classical analysis presented so far, we have explored in depth the properties of the metapotential surface in the classical limit. It is natural to ask, how do the surface properties inform the quantum behaviour of the oscillator? In this section, we present a semi-classical phase diagram of the squeeze-driven Kerr oscillator, to count the number of excited states that have sunk under the barrier. Following the Einstein-Brillouin-Keller method [70], which generalizes the notion of Bohr orbits, we quantize the action enclosed in the well below the height of the barrier and obtain the number of in-well excited states. The picture from Bohr quantization will be useful to understand the excited state spectrum and tunneling properties that we measure in the experiment. Another semiclassical result that we will present is a WKB calculation of the tunnel splitting in the quantum double-well system modeling the SKO. This WKB analysis of the SKO is an adaptation of the result of Marthaler and Dykman [102, 103], which is itself an extension of the WKB calculation of the tunnel splitting between the lowest eigenstates of an ordinary double well potential.

Action Quantization via Einstein-Brillouin-Keller (EBK) Method

In this section, we present the semiclassical method of obtaining the number of in-well states via action quantization, following the Einstein-Brillouin-Keller method, which generalizes the notion of Bohr orbits. The main result of this analysis is Fig. 3.4, a semiclassical period doubling phase diagram, where we plot contours of constant number of well-states obtained by quantizing the action enclosed per well. The results of the calculation are summarized in Table 3.1.

Properties of the Metapotential Surface for $\gamma/K \ll 1$

In the table below, we examine the properties of the metapotential surface given by Eq. (3.4a) in the double-well regime. For details on the number of levels inside the well, which we obtain via action quantization following the prescription of Einstein-Brillouin-Keller (EBK) [70], see Sect. 3.1.

First, we introduce a polar-coordinate representation of Eq. (3.17), which exploits its radial symmetry, as

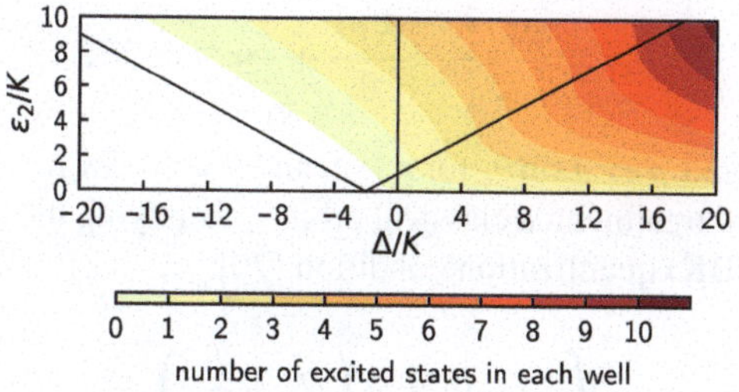

Fig. 3.4 (**a**) Semiclassical period-doubling phase diagram of the metapotential for $\gamma/K \ll 1$. Equi-state contours in parameter space are shown in color. (**b**) Characteristic metapotentials with Bohr-like orbits in black for points in the Δ/K, ϵ_2/K, where the points are marked by colored stars. In (**b**) Δ/K is kept constant at 0 and ϵ_2/K is increased. This corresponds to the condition under which the original Kerr-cat qubit was proposed [126] and operated [53, 66]

Table 3.1 Summary of metapotential properties, where the metapotential is defined by Eq. (3.4a). We recall that $\Delta_o = \omega_o - \omega_d/2$

Phase → ↓ Parameter (x, p phase space)	Double-node $-2\epsilon_2 \le \Delta_o \le 2\epsilon_2$	Triple-node $\Delta_o > 2\epsilon_2$
Area	$\frac{\Delta_o}{K} \arccos\left(\frac{-\Delta_o}{2\epsilon_2}\right) + \frac{2\epsilon_2}{K}\sqrt{1-\left(\frac{\Delta_o}{2\epsilon_2}\right)^2}$	$\frac{4\epsilon_2}{K}\sqrt{\frac{\Delta_o}{2\epsilon_2}-1} + \frac{2\Delta_o}{K}\arcsin\left(\sqrt{\frac{2\epsilon_2}{\Delta_o}}\right)$
Levels in well (#)	Area$/2\pi - 1/2$	Area$/2\pi - 1/2$
Approximation of #	$\frac{\Delta_o/K}{2} + \frac{\epsilon_2/K}{\pi} - \frac{1}{2}$	$\frac{\sqrt{8\epsilon_2\Delta_o}}{K\pi} - \frac{1}{2}$
Distance b/w nodes	$2\sqrt{\frac{\Delta_o+2\epsilon_2}{2K}}$	$2\sqrt{\frac{\Delta_o+2\epsilon_2}{2K}}$
Distance b/w saddles	0	$2\sqrt{\frac{\Delta_o-2\epsilon_2}{2K}}$
Depth of nodes	$\frac{(\Delta_o+2\epsilon_2)^2}{4K}$	$\frac{(\Delta_o+2\epsilon_2)^2}{4K}$
Depth of saddles	0	$\frac{(\Delta_o-2\epsilon_2)^2}{4K}$
Depth of barrier	$\frac{(\Delta_o+2\epsilon_2)^2}{4K}$	$\frac{2\Delta_o\epsilon_2}{K}$

$$H_{\rm cl} = \frac{\Delta_o r^2}{2} - \frac{K r^4}{4} + \epsilon_2 r^2 \cos 2\theta, \tag{3.8}$$

where $x = r\cos\theta$ and $p = r\sin\theta$, for $r \geq 0$ and $\theta \in [0, 2\pi)$.

In a semiclassical treatment, a classical orbit $\mathcal{C}_j$ satisfying the following Einstein-Brillouin-Keller (EBK) quantization condition [70]:

$$\int_{\mathcal{C}_j} dx\, dp = \hbar \left(N_j + \frac{\beta_j}{4} \right), \tag{3.9}$$

plays a special role. On the left hand side of Eq. (3.9), the action integral corresponds to the area enclosed by the contour $\mathcal{C}_j$. On the right hand side of Eq. (3.9), the non-negative integer $N_j \geq 0$ represents a quantum number and β_j is called a Maslov index; it counts the number of caustics encountered by the contour $\mathcal{C}_j$. For an orbit in the Kerr-cat metapotential, we have $\beta_j = 2$. Thus the condition in Eq. (3.9) states that only those orbits whose enclosed area satisfy a condition given by non-negative integers n_j and $\beta_j = 2$ correspond to allowed quantum orbits.

With this condition stated, one can ask a simple question: given a set of Δ, ϵ_2, how many in-well or bound states exist in the classical metapotential surface? This will be obtained by computing the number of allowed states below the separatrix, which separates bound and unbound states.

From the calculations detailed below, we find the number of bound states as

$$N \sim \begin{cases} \frac{(\Delta+2K)/K}{2} + \frac{\epsilon_2/K}{\pi} - \frac{1}{2} & -2\epsilon_2 \leq \Delta < 2\epsilon_2 \\ \frac{\sqrt{8\epsilon_2(\Delta+2K)}}{K\pi} - \frac{1}{2} & \Delta \geq 2\epsilon_2. \end{cases} \tag{3.10}$$

We demonstrate in Fig. 3.11 the value of the semi-classical action quantization condition in predicting the locality in phase space of even the excited states of the squeeze-driven Kerr oscillator.

Separatrix Area in the Double-Node Phase: $-2\epsilon_2 \leq \Delta_o < 2\epsilon_2$

In the double-node phase, the separatrix has a special name called the Bernoulli's lemniscate and its equation is given as

$$r^2 = \frac{2\Delta_o}{K} + \frac{4\epsilon_2}{K}\cos 2\theta, \tag{3.11}$$

and $-\theta_c \leq \theta \leq \theta_c$, where $\theta_c = \frac{1}{2}\arccos\frac{-\Delta_o}{2\epsilon_2}$. We compute the area of a half the lemniscate as

$$
\begin{aligned}
\int_{C_j} dx\, p &= \frac{1}{2}\int_{-\theta_c}^{\theta_c} d\theta\, r^2 \\
&= \int_0^{\theta_c} d\theta\, \frac{2\Delta_o}{K} + \frac{4\epsilon_2}{K}\cos 2\theta \\
&= \frac{\Delta_o}{K}\arccos\left(\frac{-\Delta_o}{2\epsilon_2}\right) + \frac{2\epsilon_2}{K}\sqrt{1-\left(\frac{\Delta_o}{2\epsilon_2}\right)^2} \\
&\sim \frac{\Delta_o}{K}\left(\frac{\pi}{2}+\frac{\Delta_o}{2\epsilon_2}\right) + \frac{2\epsilon_2}{K}\left(1-\frac{1}{2}\left(\frac{\Delta_o}{2\epsilon_2}\right)^2\right) \qquad |\Delta_o/2\epsilon_2| \ll 1 \\
&= \frac{\pi}{2}\frac{\Delta_o}{K} + 2\frac{\epsilon_2}{K} \\
&= \pi + \frac{\pi}{2}\frac{\Delta}{K} + 2\frac{\epsilon_2}{K}.
\end{aligned} \tag{3.12}
$$

Note that for $\Delta = 0$, Eq. (3.12) reduces to $\pi + 2\epsilon_2/K$.

Separatrix Area in the Triple-Node Phase: $\Delta_o \geq 2\epsilon_2$

The separatrix in the triple-node phase is given as

$$
r_\pm^2 = \frac{\Delta_o}{K} + \frac{2\epsilon_2}{K}\cos 2\theta \pm \frac{4\epsilon_2\cos\theta}{K}\sqrt{\frac{\Delta_o}{2\epsilon_2} - \sin^2\theta} \tag{3.13}
$$

and $-\theta_c \leq \theta \leq \theta_c$, where $\theta_c = \frac{\pi}{2}$. When plotted, this separatrix carves a bean-like shape.

Remarkably, we find an exact analytic expression for the area of this surface as

$$
\begin{aligned}
\int_{C_j} dx\, p &= \frac{1}{2}\int_{-\theta_c}^{\theta_c} d\theta\, (r_+^2 - r_-^2) \\
&= \int_{-\pi/2}^{\pi/2} d\theta\, \frac{4\epsilon_2\cos\theta}{K}\sqrt{\frac{\Delta_o}{2\epsilon_2} - \sin^2\theta} = \frac{4\epsilon_2}{K}\int_0^1 dt\, \sqrt{\frac{\Delta_o}{2\epsilon_2} - t^2} \\
&= \frac{4\epsilon_2}{K}\sqrt{\frac{\Delta_o}{2\epsilon_2} - 1} + \frac{2\Delta_o}{K}\arcsin\left(\sqrt{\frac{2\epsilon_2}{\Delta_o}}\right) \\
&\sim \frac{2\sqrt{8\epsilon_2\Delta_o}}{K}, \qquad \Delta_o/2\epsilon_2 \gg 1 \\
&\sim \frac{2\sqrt{8\epsilon_2(\Delta + 2K)}}{K}, \qquad \Delta_o/2\epsilon_2 \gg 1.
\end{aligned} \tag{3.14}
$$

Indeed the effective Hamiltonian of the squeezed Kerr oscillator (SKO) Eq. (3.1) can be diagonalized in its operator-valued form. In the following section, we will examine in detail the spectrum of the SKO as a function of Δ/K and ϵ_2/K, the only free parameters of the Hamiltonian. We will then examine the wave functions of the SKO in this parameter space to better understand their structure and tunneling properties.

Operator Space Representation

Spectral Kissing as a Function of ϵ_2

We present the spectrum of Eq. (3.1) in Fig. 3.6a as a function of the squeezing drive amplitude ϵ_2/K for $\Delta = 0$. The eigenstates of effective Hamiltonian of the SKO Eq. (3.1) have well-defined parity. For $\epsilon_2/K = 0$, the Fock states which have well-defined parity are the eigenstates. Since the squeezing interaction is itself parity-conserving, when $\epsilon_2 \neq 0$, only states with the same parity are coupled. In Fig. 3.6a, we color the energy levels by their parity, and further reference them to the ground state energy, which due to the choice of the rotating frame at $\omega_d/2$, happens to be the highest energy eigenstate. For $\epsilon_2/K \to 0$, the spectrum corresponds to the expected Kerr anharmonic spectrum of the SNAIL circuit oscillator. As the squeezing interaction ϵ_2 grows, the spectrum becomes pairwise degenerate with levels of different photon-number parity approaching each other in an exponential fashion. We refer to this exponential approach of energy levels as *spectral kissing*.

We provide an intuitive way to understand the spectrum following the semiclassical analysis introduced in Sect. 3.1. For $\Delta = 0$, according to the action quantization result Eq. (3.10), the excited state manifold N gets ensconced inside the quantum double well, whenever $\epsilon_2/K = \left(N + \frac{1}{2}\right)\pi$ for $\Delta = 0$. As the squeezing amplitude ϵ_2 increases, excited states are ensconced into the metapotential well and, coupled only by quantum tunneling, exponential level kissing takes place. The second X-axis in Fig. 3.5a corresponds to the number of quantized bound states obtained by applying this semiclassical condition. This simple argument gets more nuanced in the presence of dissipation, but plays an important role in understanding the decoherence properties of the SKO as a function of ϵ_2. Both the spectral kissing and the decoherence properties of this quantum double well system were measured for the first time in [53].

Multi-Level Degeneracies as a Function of Δ

We present the spectrum of Fig. 3.1 in Fig. 3.6b as a function of the squeezing drive detuning Δ/K for $\epsilon_2/K = 2$. Remarkably, the oscillator experiences multi-level degeneracies in the spectrum: when the detuning Δ of the drive from the oscillator frequency equals an even multiple of the Kerr coefficient K, $\Delta/K =$

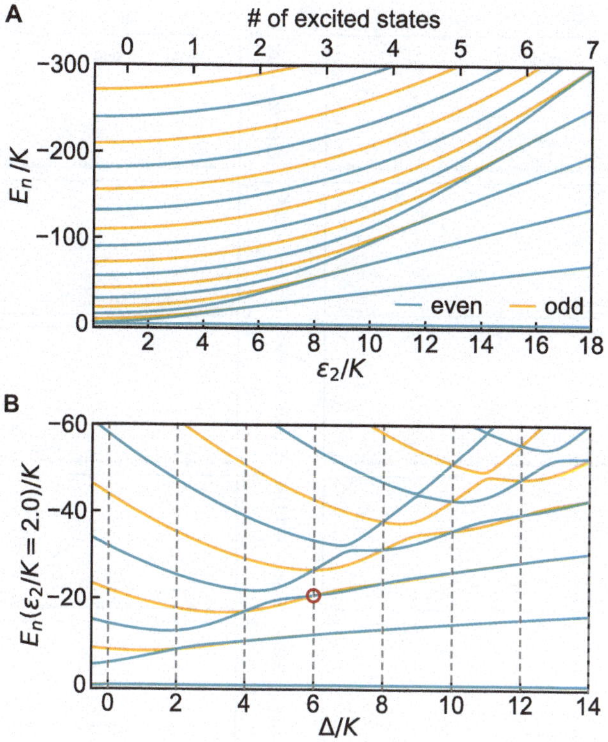

Fig. 3.5 Spectrum of Eq. (3.1) versus ϵ_2 in panel (**a**) shows pairwise kissing and vesus Δ for $\epsilon_2/K = 2$ in panel (**b**) shows multilevel degeneracies. The spectrum is colored by parity. In (**b**) dashed grey lines correspond to even integer values of Δ/K. The red circle marks an exact degeneracy between an odd and an even parity state

$2m$, the oscillator experiences $m + 1$ exact spectral degeneracies. Not only can these degeneracies be turned on-and-off on demand, but their number is tunable. Importantly, these degeneracies are robust as they are completely independent of the drive amplitude ϵ_2/K.

Hidden symmetries typically manifest as unexpected, exact spectral degeneracies. Establishing the connection between symmetries and degeneracies in a system is of fundamental importance for the understanding and control of its structural and dynamical properties. Unexpectedly, the SKO has multiple tunable spectral degeneracies that always occur when $\Delta/K = 2m$, and are independent of ϵ_2. An open question is what is the tunable hidden symmetry underlying these degeneracies? An obvious, yet incorrect answer is parity symmetry. We discuss this important property next.

The SKO has the remarkable property: for $\Delta/K = 2m$, the first $m + 1$ pairs of levels become decoupled from the rest of the oscillator's Hilbert space. Their eigenenergies and eigenstates become exactly solvable and present $m + 1$ robust

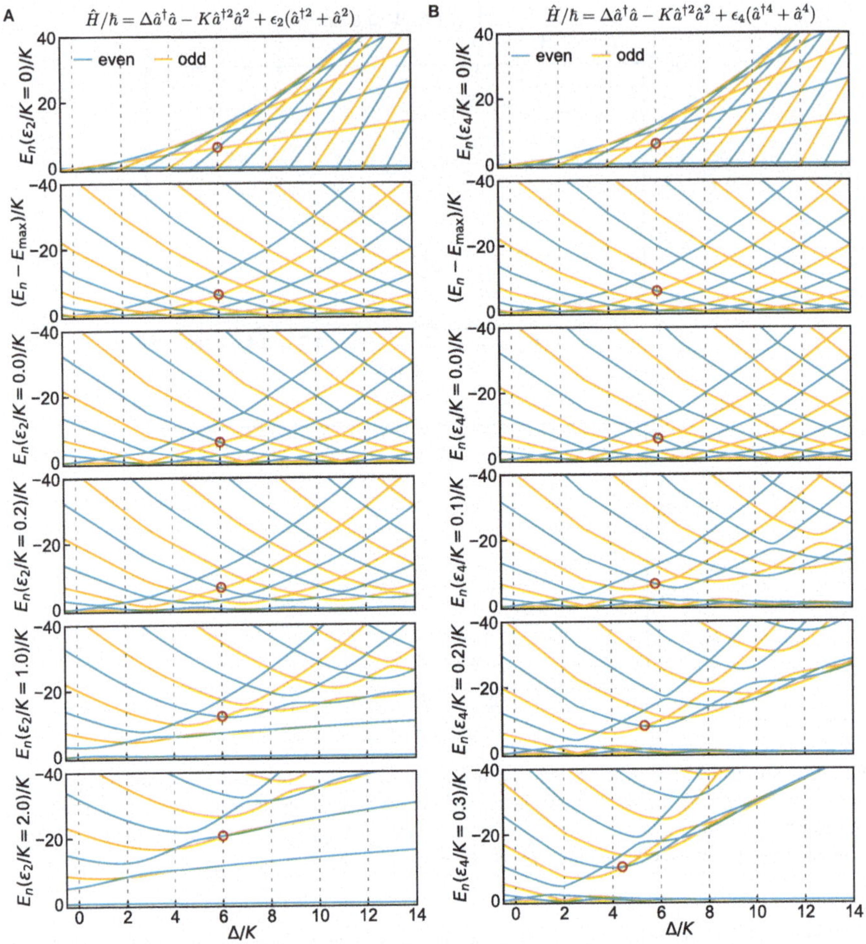

Fig. 3.6 Robust spectral degeneracies and parity symmetry. Spectrum of (**a**) the squeeze-driven Kerr oscillator, with Eq. (3.1), and (**b**) a different member of the Kerr parametric oscillator family with parity symmetry, with $\hat{H} = \Delta\hat{a}^\dagger\hat{a} - K\hat{a}^{\dagger 2}\hat{a}^2 + \epsilon_4(\hat{a}^{\dagger 4} + \hat{a}^4)$, as a function of Δ/K for different values of ϵ_2/K and ϵ_4/K respectively. Dashed lines mark Δ/K corresponding to even integers. Left panel indicates that even for non-perturbative values of ϵ_2/K, the locations of crossings of even (blue) and odd (orange) parity eigenstates occur at even values of Δ/K. Right panel indicates that even for the parity preserving perturbation controlled by ϵ_4/K, the locations of the crossings of even and odd parity states get renormalized. Red circle tracks one such crossing

degeneracies in between states of different photon-number parity. Critically, note that the resonance condition for these degeneracies is independent of the value of the squeezing drive amplitude ϵ_2.

First, to show this, we begin by considering the squeezing drive as a perturbation to the Kerr oscillator described by the Hamiltonian $\hat{H}_K/\hbar = \Delta\hat{a}^\dagger\hat{a} - K\hat{a}^{\dagger 2}\hat{a}^2$ which is exactly solvable: its eigenstates are Fock states $|n\rangle$ and their energies are

$E_n^{(0)} = \Delta n - Kn(n-1)$, which, as a function of Δ, are lines with integer slope that we plot in the top row of Fig. 3.6a. The even(n)-odd(l) degeneracies read $E_n = E_l$ and imply $\Delta/K = 2m$ where $m = (n+l-1)/2 \geq 0$ is any nonnegative integer. In the second row, we plot the transition spectrum with respect to the ground state at $\epsilon_2 = 0$, which, due to the choice of rotating frame, corresponds to the *highest* energy eigenstate. This is the directly experimentally observable *transition* spectrum from the *ground* state. We further note that the ground state changes with Δ; remarkably, for $\epsilon_2 = 0$, at $\Delta/K = 2m$, the ground state is $(|m\rangle + |m+1\rangle)/\sqrt{2}$. This special property of the squeeze-driven Kerr oscillator has technological applications [176]. In the following rows, we plot the transition spectrum for increasing values of squeezing drive amplitude ϵ_2.

Indeed, it is clear that the squeezing drive renormalizes the energies of the Kerr oscillator. Level crossings of the Kerr oscillator with different parity remain exact crossings in the presence of the squeezing drive, since the interaction preserves parity. However, the remarkable feature is that these crossings are locked to where Δ equals an even multiple of K. In the following text, we justify this property, first via a perturbative and then provide a to-all-order proof.

Perturbative Analysis of Degeneracies

To first order in perturbation theory, we see that this even and odd Fock states remain decoupled (energy level crossings) under the parity conserving squeezing drive: $E_n^{(1)} = \langle n|(\hat{a}^{\dagger 2} + \hat{a}^2)|n+1\rangle = 0$. The condition for crossings of consecutive levels with different parity ($E_n = E_{n+2}$) reads instead $\Delta/K = (2n+1)$. To first order in perturbation theory, the avoided crossing amplitude is $E_n^{(1)} = \epsilon_2\sqrt{(n+1)(n+2)}$.

As a next approximation to the problem, we see the robustness of the crossings of consecutive levels with different parity (at $\Delta = 2nK$) by computing the second order correction to the nth energy levels $E_n^{(2)}$ and comparing it to the correction for the $(n+1)$th energy level

$$E_{n+1}^{(2)} = \epsilon_2^2\left(\frac{(n+3)(n+2)}{-2\Delta + 2K(2n+3)} + \frac{(n+1)n}{2\Delta - 2K(2n+1)}\right),$$

to find that $E_n^{(2)} = E_{n+1}^{(2)}$ for $\Delta/K = 2n$. This robustness can be seen in Fig. 3.6a (all panels), where we see that the crossing shifts in energy but remains locked to Δ/K equal to even non-negative integers. The perturbation theory argument is easily generalized to non-consecutive level crossings and anti-crossings to this order. A similar perturbative argument was made in [176].

Non-perturbative Analysis of Degeneracies

To prove that the location of the degeneracies in Δ is independent of the squeezing drive amplitude to all orders we observe that we can write the Hamiltonian in Eq. (3.1) as

$$\hat{H} = \lambda_1(\hat{a}^{\dagger 2} - \alpha^2)(\hat{a}^2 - \alpha^2) + \lambda_2(\hat{a}^2 - \alpha^2)(\hat{a}^{\dagger 2} - \alpha^2), \tag{3.15}$$

where, for $\Delta/K = 2m$ (m non-negative integer), we have $\lambda_1 = -K(1 + m/2)$, $\lambda_2 = mK/2$ and $\alpha = \pm\sqrt{\epsilon_2/K}$, and which is a generalization of the factorization condition proposed in [126] for $\Delta = 0$. We next consider the displaced Hamiltonian $\hat{H}^+ = \hat{D}(+\alpha)\hat{H}\hat{D}^\dagger(+\alpha)$, which brings one of the wells to the origin of phase space. In this frame, the Hamiltonian operator can be written as[3]

$$\begin{aligned}\hat{H}^+ =& - K\left(\hat{a}^{\dagger 2}\hat{a}^2 + (4\alpha^2 + 2m)\hat{a}^\dagger\hat{a}\right)\\ &- 2K\alpha[\hat{a}^\dagger\hat{a} - (m+1)]\hat{a}^\dagger\\ &- 2K\alpha[\hat{a}^\dagger\hat{a} - m]\hat{a}.\end{aligned}$$

While the first line is number conserving, the next two lines couple only consecutive Fock states. In matrix form, it is tridiagonal in the Fock basis $|n\rangle$. By examining the square brackets in the above expression, we see that the off-diagonal elements are exactly zero for $n = m$ and $n = m + 1$. Thus, the first $m+1$ states decouple from the rest of the oscillator's Hilbert space. The finite matrix is Hermitian, negative-semidefinite, and tridiagonal so it is exactly diagonalizable. Finally, we note that in phase space, a displacement of the metapotential surface, which is mirror-symmetric about $x = 0$, is identical to an opposite displacement composed with a rotation of 180° around the origin. Since the photon-number parity operator $\hat{\Pi} = e^{i\pi\hat{a}^\dagger\hat{a}}$ commutes with the Hamiltonian ($[\hat{\Pi}, \hat{H}] = 0$) the rotation is a symmetry of the system. Specifically; $\hat{H}^- = \hat{D}(-\alpha)\hat{H}\hat{D}^\dagger(-\alpha) \Rightarrow \hat{\Pi}\hat{H}^-\hat{\Pi} = \hat{H}^+$. We thus have two sets of equivalent[4] $m + 1$ exactly solvable eigenenergies, and $2(m + 1)$ linearly independent equations,[5] which imply the

[3] Note that, without specializing Δ, one can directly write from Eq. (3.1) in the main text, or equivalently from Eq. (3.15): $\hat{H}^+ = -K\left(\hat{a}^{\dagger 2}\hat{a}^2 + (4\alpha^2 + \Delta/K)\hat{a}^\dagger\hat{a}\right) - 2K\alpha[\hat{a}^\dagger\hat{a} - (\Delta/2K + 1)]\hat{a}^\dagger - 2K\alpha[\hat{a}^\dagger\hat{a} - \Delta/2K]\hat{a}$. From this expression one can directly derive the sub-space decoupling condition to be $\Delta/K = 2m$, in an exact manner, without relying in perturbative calculation or any previous knowledge existence of the resonance. The independence of the sub-space decoupling condition with respect to ϵ_2 is explicit.

[4] Note that for the off-diagonal elements the parity transformation produces a minus sign ($\hat{\Pi}|n\rangle\langle n\pm 1|\hat{\Pi} = -|n\rangle\langle n \pm 1|$) that manifests in $\alpha \to -\alpha$: $H^+_{n,n\pm1} = -H^-_{n,n\pm1}$. This leaves the finite characteristic polynomial invariant.

[5] The elements of the finite set of eigenvectors of $\hat{H}^+$, $\{|\phi^+_{k\leq m+1}\rangle\}_k$, are linearly independent from the finite set of eigenvectors of $\hat{H}^-$, $\{|\phi^-_{k\leq m+1}\rangle\}_k$, since they are spanned by the first

existence of $m+1$ degeneracies in the spectrum for $\Delta/K = 2m$. The $2(m+1)$ eigenstates $|\psi^{\pm}_{k\leq m+1}\rangle = \hat{D}(\pm\alpha)|\phi^{\pm}_{k\leq m+1}\rangle$ of $\hat{H}$, where $|\phi^{-}_{k\leq m+1}\rangle = \hat{\Pi}|\phi^{+}_{k\leq m+1}\rangle$ and $\hat{H}^{+}|\phi^{+}_{k\leq m+1}\rangle = E_{k\leq m+1}|\phi^{+}_{k\leq m+1}\rangle$, found in this way are not orthogonal, but thanks to the two-fold degeneracy condition we can take the superposition of the right (+) and left (−) kth displaced state to get an orthogonal basis in each of the $m+1$ two-fold degenerate sub-spaces: $|\mathcal{C}^{\pm}_{k\leq m+1}\rangle \propto D(+\alpha)|\phi^{+}_{k\leq m+1}\rangle \pm D(-\alpha)|\phi^{-}_{k\leq m+1}\rangle$. These $2(m+1)$ pairwise-degenerate eigenstates of energy are also eigenstates of parity.[6] In this work we name these pairs of degenerate states the Δ-cats.

Note, that the robustness of the resonance condition is a peculiar symmetry property of the squeeze-driven Kerr oscillator and not a property of generic Kerr parametric oscillators. The existence of this robust degeneracies begs the question: what are the hidden symmetries associated with these degeneracies, if any? We show in Fig. 3.6b, as an example, the spectrum of $\hat{H} = \Delta\hat{a}^{\dagger}\hat{a} - K\hat{a}^{\dagger 2}\hat{a}^{2} + \epsilon_4(\hat{a}^{\dagger 4} + \hat{a}^{4})$, where the location in Δ of the super-parity resonances depend on the value of the parametric drive amplitude ϵ_4. Note, also, that even if the multilevel resonances in Fig. 3.6b are displaced with the value of the parametric drive amplitude (red circles), they are locked together to a running resonance condition: the point of exact solvability is changed by the drive. The phenomenon corresponds to deep symmetries [79, 82] of these type of, as of now, engineerable bosonic Hamiltonians and will be discussed in detail in a separate publication.

To provide greater insight into these spectral degeneracies, we examine the Wigner functions and the wave functions of the associated ground and excited states next.

Oscillations in the Classically Forbidden Region in the Localized Wavefunctions of the SKO

The SKO is modeled by a quantum double-well in phase space wherein the tunnelling of the ground and excited states can simultaneously be nulled, at finite barrier height, due to an unexpected destructive interference of tunneling paths in the classically forbidden region. This phenomenology was discovered for the ground

$m+1$ displaced Fock states in different directions (±, see text). Indeed, any Fock state i that is displaced has support in all (undisplaced) Fock states j's: for $j > i$ the formula reads $|\langle j|\hat{D}(\pm 2\alpha)|i\rangle| = \left(\frac{i!}{j!}\right)^{1/2} |2\alpha|^{j-i} e^{-2|\alpha|^2} |L_i^{(j-i)}(4|\alpha|^2)| > 0$, where $L_i^{(j-i)}$ is an associated Laguerre polynomial (note that the matrix element tends to zero rapidly as $|\alpha|$, or equivalently $|\epsilon_2|$, grows, while the decoupled subspace condition, and ultimately the proof itself, is independent of these values for as long as they are non zero. If $\epsilon_2 = 0$, the proof is trivial and is given in the previous subsection). In other words, the linear independence manifests here explicitly in that $|\phi^{\pm}_{k\leq m+1}\rangle$ have no defined parity, yet $|\phi^{+}_{k\leq m+1}\rangle = \hat{\Pi}|\phi^{-}_{k\leq m+1}\rangle$ (note that $[\hat{\Pi}, \hat{H}^{\pm}] \neq 0$). Ultimately, $|\langle\phi^{-}_{k\leq m+1}|\hat{D}^{\dagger}(-\alpha)\hat{D}(+\alpha)|\phi^{+}_{k\leq m+1}\rangle| = |\langle\phi^{+}_{k\leq m+1}|\hat{D}(2\alpha)|\phi^{+}_{k\leq m+1}\rangle| < 1$ if $|\alpha| > 0$.

[6] Specifically $\hat{\Pi}|\mathcal{C}^{\pm}_{k\leq m+1}\rangle = \pm|\mathcal{C}^{\pm}_{k\leq m+1}\rangle$ and are thus orthogonal. We used $\hat{D}(+\alpha)\hat{\Pi} = \hat{\Pi}\hat{D}(-\alpha)$.

state manifold in the double-node phase by [102, 103] using a semiclassical WKB analysis.

Before looking at this phase-space double well model system, let us recap the tunneling properties of the ordinary double-well potential. We define the ordinary double-well model system as:

$$H = \frac{p^2}{2} + V(x), \qquad \text{with} \qquad V(x) = -\frac{k_2}{2}x^2 + \frac{k_4}{4}x^4,$$

where $k_2,\ k_4 > 0$. This potential has a saddle at $x_s = 0$, with $V(x_s) = 0$ and nodes at $x_n = \pm\sqrt{k_2/k_4}$ with the left and right well depth given by $V(x_n) = -k_2^2/(4k_4)$. The barrier height is given by $V(x_n) - V(x_s) = k_2^2/(4k_4)$.

The study of tunneling usually begins by considering a localized wave packet in one well, which is written as the superposition of the wavefunctions of the two lowest laying energy states ψ_+ and ψ_-.[7] Their energy difference is denoted by $\delta E = E^+ - E^-$ and the left- and right-localized wavefunctions read

$$\psi_l = \frac{\psi_+ + \psi_-}{\sqrt{2}} \qquad \psi_r = \frac{\psi_+ - \psi_-}{\sqrt{2}}. \tag{3.16}$$

On the left column of Fig. 3.7, we plot the left and right-localized wavefunctions in red and blue respectively for a $k_2 = 3,\ k_4 = 1$, b $k_2 = 2,\ k_4 = 1$, and b $k_2 = 4,\ k_4 = 2$ respectively. The wavefunctions are computed by numerical diagonalization of the Hamiltonian. In the classically forbidden region, as one should expect, the wavefunctions display evanescent decay [64].

In the right column of Fig. 3.7, we contrast the exact localized wavefunctions of the ordinary double-well potential with those of the squeeze-driven Kerr oscillator, both of which have been obtained numerically. The parameters Δ, K, and ϵ_2 were chosen so that a cut of the effective Hamiltonian surface at $p = 0$ yields an identical double-well potential as the left column. The wavefunctions of the full squeeze-driven Kerr oscillator are computed numerically. Importantly, b, d, and f show the localized wavefunctions for $\Delta/K = 2$, 1, and 0 respectively, corresponding to the coherent destructive, constructive, and again destructive interference. Interestingly, in the classically forbidden region, in b and d, oscillations accompany decay in the wavefunction [102, 103]. This is due to the underlying driven nature of our system, providing a quartic term in momentum, which here reflects in the oscillatory nature of the wavefucntions in the classically forbidden region (see next page).

[7] From a perturbation theory point of view this corresponds to the bonding and anti-bonding of the decoupled well states [95]. The zero point energy of the individual wells, in the absence of tunneling, is $E_0 = \sqrt{k_2}/2$. In the presence of tunneling the system's energies can be approximated by $E^{\pm} = E_0 \pm \delta E$.

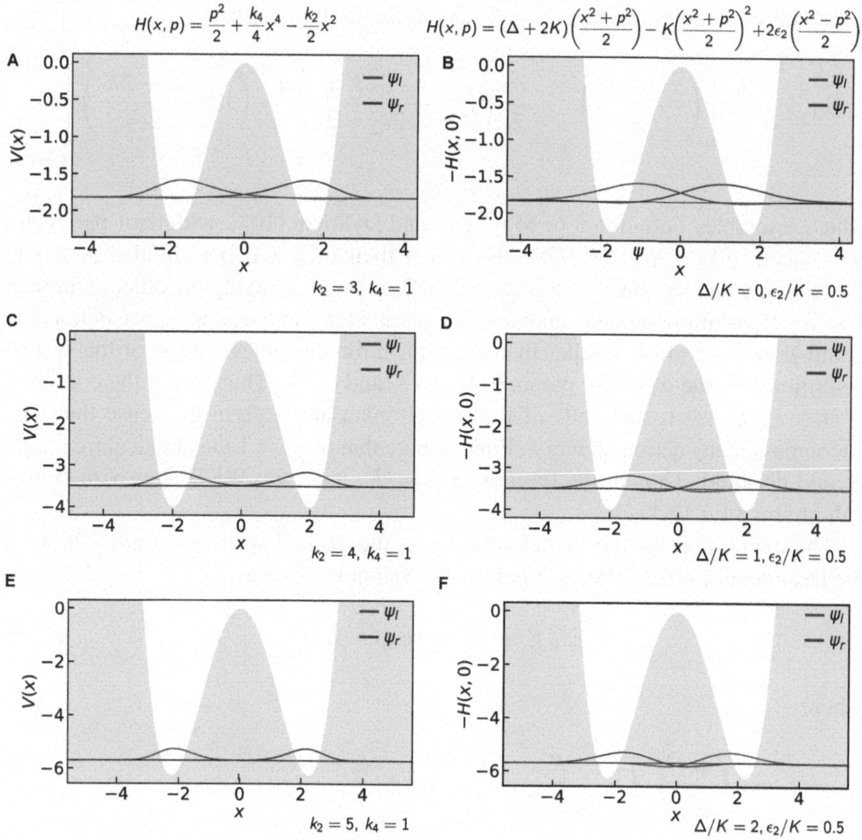

Fig. 3.7 Localized position wavefunctions of the ground state manifold for (**a**), (**c**), (**e**) an ordinary double well potential and (**b**), (**d**), (**f**) for a squeeze-driven Kerr oscillator. The Hamiltonian parameters in (**a**), (**c**), and (**e**) have been chosen to produce a double-well with the same depth and the well separation as those of (**b**), (**d**), and (**f**) respectively. The value of Δ/K is chosen to be (**b**) $\Delta/K = 2$, (**d**) $\Delta/K = 1$, and (**f**) $\Delta/K = 0$ corresponding to the destructive, constructive, and destructive interference of tunneling respectively in Eq. (3.4a). In the right panel, oscillations accompany decay of the wavefunction in the classically forbidden region, marked in grey. In the left panel, the wavefunction exhibits pure decay in the classically forbidden region. The cancellation of the tunnel-splitting at even multiples of K in Fig. 3.5b can be understood as the destructive interference of the wavefunction in the classically forbidden region of the squeeze-driven Kerr oscillator. In [103], Marthaler and Dykman found an analytical expression for the WKB tunnel splitting of the ground state manifold. See Chap. 5 for comparisons of the extracted tunnel splitting from experiment with the predictions from the WKB calculation

WKB Approximation of Tunnel Splitting for the Ground State Manifold of the SKO

Marthaler and Dykman predicted this oscillatory behaviour of the wave functions of a parametrically modulated oscillator in the classically forbidden region. To connect our result with theirs, we take $\lambda = K/2\epsilon_2$ in Eq. (3.4a) to obtain

$$H_{\rm cl} = -\frac{K}{\lambda^2}\left[\left(\frac{x^2+p^2}{2}\right)^2 - \frac{x^2}{2}\left(1+\frac{\Delta+2K}{2\epsilon_2}\right) + \frac{p^2}{2}\left(1-\frac{\Delta+2K}{2\epsilon_2}\right)\right], \tag{3.17}$$

which resembles Equation 5 of Marthaler and Dykman [103], with their parameter μ is taken to be $(\Delta + 2K)/2\epsilon_2$. Note that their effective Hamiltonian seems to contain only one parameter $\mu = (\Delta+2K)/2\epsilon_2$. But in actuality the other parameter $\lambda = K/2\epsilon_2$ is implicit in the analysis. The parameter λ in the phase space defined by x and p must be much smaller than the typical dimensionless action of the system determined by the well-size parameters: Δ/K and ϵ_2/K. Thus, under the condition $\Delta/K,\ \epsilon_2/K \gg 1$, the wells of the Hamiltonian are large in the sense that they encompass many action quanta λ. Finally, note that for $\lambda \approx 1$ the classical treatment should not hold. Thus in the limit $\Delta/K,\ \epsilon_2/K \gg 1$, the WKB approximation is valid to treat Eq. (3.1).

The expression for the tunnel splitting in the ground state manifold, following the treatment of [102, 103], applied to our system is given as

$$\delta E = f\cos\theta\exp(-A), \tag{3.18}$$

where

$$\begin{aligned} f &= 2\left(\frac{4\epsilon_2}{K}\right)^2\left(\frac{K}{\pi(\Delta+2K)}\right)^{1/2}\left(1+\frac{\Delta+2K}{2\epsilon_2}\right)^{5/4} \\ \theta &= \frac{\pi}{2}\left(\frac{\Delta+2K}{K}-1\right) \\ A &= \frac{2\epsilon_2}{K}\left(\frac{\Delta+2K}{2\epsilon_2}+1\right)^{1/2} \\ &\quad - \frac{\Delta+2K}{K}\log\left(\left(\frac{2\epsilon_2}{\Delta+2K}\right)^{1/2}+\left(1+\frac{2\epsilon_2}{\Delta+2K}\right)^{1/2}\right), \end{aligned} \tag{3.19}$$

where, the above expression is only valid for $\Delta/K \gg 1$.

As a function of Δ, δE from Eq. (3.18) is nulled whenever the cosine argument is an odd multiple of $\pi/2$ and this corresponds to the condition $\frac{\Delta}{K} = 2m$, for $m \geq 0$. Beyond the tunneling nulls under this condition, the $\exp(-A)$ factor yields an exponential suppression of tunneling as a function of both Δ and ϵ_2. We examine the tunneling properties in the ground state manifold as a function of Δ and ϵ_2 in Fig. 3.8c and d.

There are two failure modes for the WKB approximation. The first condition corresponds to the condition $\Delta/K \lesssim 1$, and the other corresponds to the condition $\epsilon_2/K \ll 1$. In Fig. 3.8, we compare the tunnel splitting between the lowest

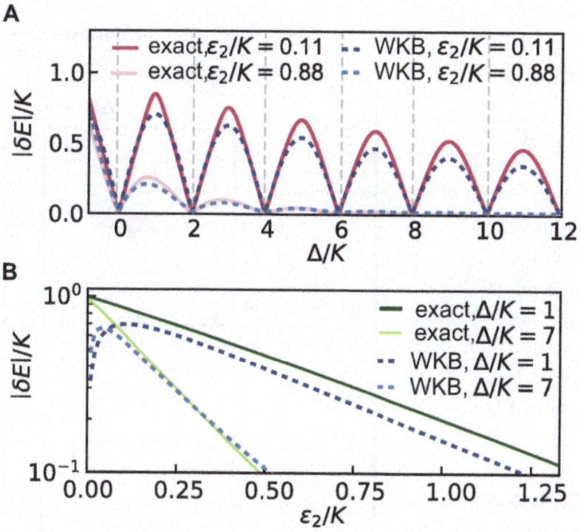

Fig. 3.8 Comparing two different theoretical models' predictions of the tunnel-splitting between the lowest pair of eigenstates of the SKO: WKB analysis (see Sect. 3.1) and exact numerical diagonalization of Eq. (3.1). (**a**) Tunnel-splitting as a function of Δ/K and (**b**) as a function of ϵ_2/K. In (**a**) solid pink lines represent the result from exact numerical diagonalization and dashed blue lines represent the WKB result presented in Sect. 3.1, which is itself adapted from [103]. Both calculations indicate that the ground-state tunnel-splitting is nulled when $\Delta/K = 2m$ and that increasing Δ/K further lowers the tunnel-splitting. The condition $\Delta/K = 2m$ corresponds to the condition for the destructive interference of tunnel-paths (see Fig. 1.2), whereas $\Delta/K = 2m + 1$ corresponds to their constructive interference. By comparing the light pink line with the dark pink line, it is clear that increasing ϵ_2/K for fixed Δ causes a further reduction in the tunneling. This is better demonstrated in (**b**) where the tunnel-splitting is plotted as a function of ϵ_2/K keeping the value of Δ/K fixed to conditions of the constructive interference of tunneling. This plot shows that the tunnel-splitting is suppressed exponentially in ϵ_2/K, and larger the value of Δ/K, faster the tunneling suppression. (**a**) and (**b**) also demonstrate the remarkable agreement between WKB analysis and exact diagonalization for $\Delta/K \gtrsim 1$ and $\epsilon_2/K \gtrsim 1$. Note that the spectral degeneracies in the excited state correspond to the extension of this WKB argument to the excited state manifold

eigenstates obtained by an exact diagonalization of Eq. (3.1) with the semiclassical result Eq. (3.18). Note that WKB works remarkably well even outside its domain of validity ($\epsilon_2/K < 1$). In [154], we measured this tunnel splitting for the ground state manifold, and compared the experimentally extracted tunnel splitting with the results of [103]. The comparison between our experiment and this theory is presented in Chap. 5.

Wigner Functions of Ground and Excited State Manifolds

Not only in a semiclassical picture, but even in a quantum picture, increasing Δ/K and ϵ_2/K yields states that are increasingly localized in phase space, thereby validating the semiclassical picture introduced here. The Wigner functions of the

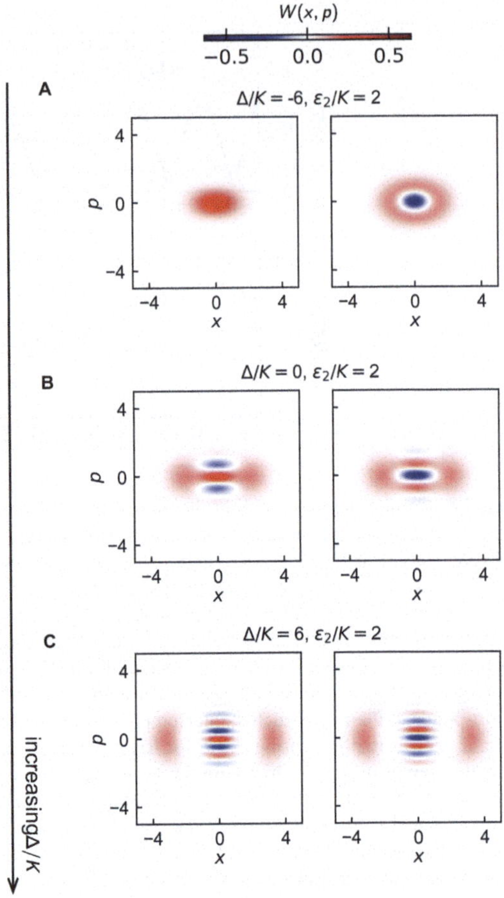

Fig. 3.9 Ground state Wigner functions of the SKO for increasing values of Δ/K and fixed ϵ_2/K. In (**a**) $\Delta/K = -6$, (**b**) $\Delta/K = 0$, and (**c**) $\Delta/K = 6$ respectively. When $\Delta/K \ll 0$, the eigenstates are squeezed Fock states. For $\Delta/K > 0$, increasing Δ/K yields Schrödinger cat states with increasing photon number. Following Table 3.1, the mean photon number of the cat state is given by $\Delta/(2K) + \epsilon_2/K$. Thus, panel (**b**) corresponds to a 2 photon cat and panel (**c**) corresponds to a 5 photon cat

lowest eigen-manifold of Eq. (3.24) is plotted in Fig. 3.9 while the excited state manifold of Eq. (3.24) is plotted in Fig. 3.10. These plots indeed confirm the semiclassical picture introduced with action quantization in Sect. 3.1. Specifically, increasing Δ and ϵ_2 yields the fastest rate of ensconcement of excited states into the double-well barrier (Fig. 3.11).

Literature Survey: The Relationships Between Different Squeeze-Driven Kerr Oscillator Models

Before we present our experimental results, we survey different SKO models treated in the literature and examine the relationships between them.

In 1993, Wielinga and Milburn [161] proposed a quantum optical model that they called *the dynamical equivalent of the double-well potential*. The interest of the

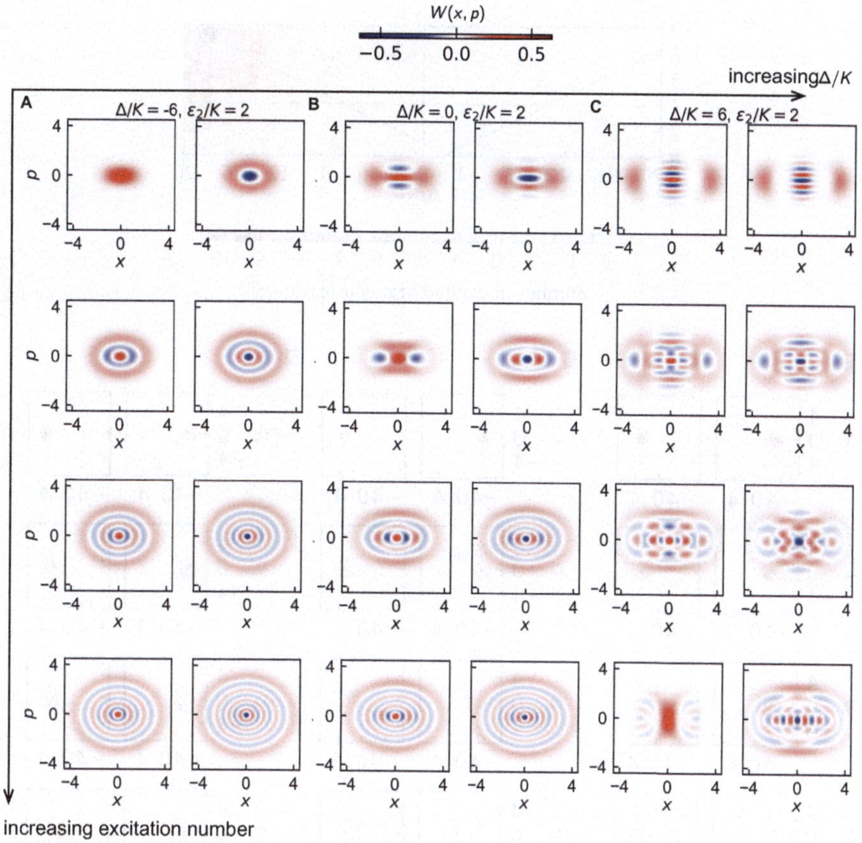

Fig. 3.10 Ground and excited states; Wigner functions of the SKO Eq. (3.1) for $\epsilon_2/K = 2$ and **(a)** $\Delta/K = -6$, **(b)** $\Delta/K = 0$, and **(c)** $\Delta/K = 6$ respectively. To highlight the ground state Wigner functions, we isolated the top row of this plot and presented it in Fig. 3.9. Note that the Wigner functions of higher excited eigenstates become increasingly wrinkled and delocalized in phase space.

problem, to them, was that their model exhibited a double-well structure in classical phase space, and quantum mechanical ground state tunneling between them. The Hamiltonian they addressed is

$$\hat{H}_{\text{WM}} = -K(\hat{a}^\dagger \hat{a})^2 + \epsilon_2(\hat{a}^{\dagger 2} + \hat{a}^2). \tag{3.20}$$

In 2017, the theoretical discovery of the Kerr-cat qubit by Puri et al. [126] relied on the fact that the ground states of

$$\hat{H}_{\text{PBB}} = -K\hat{a}^{\dagger 2}\hat{a}^2 + \epsilon_2(\hat{a}^{\dagger 2} + \hat{a}^2) \tag{3.21}$$

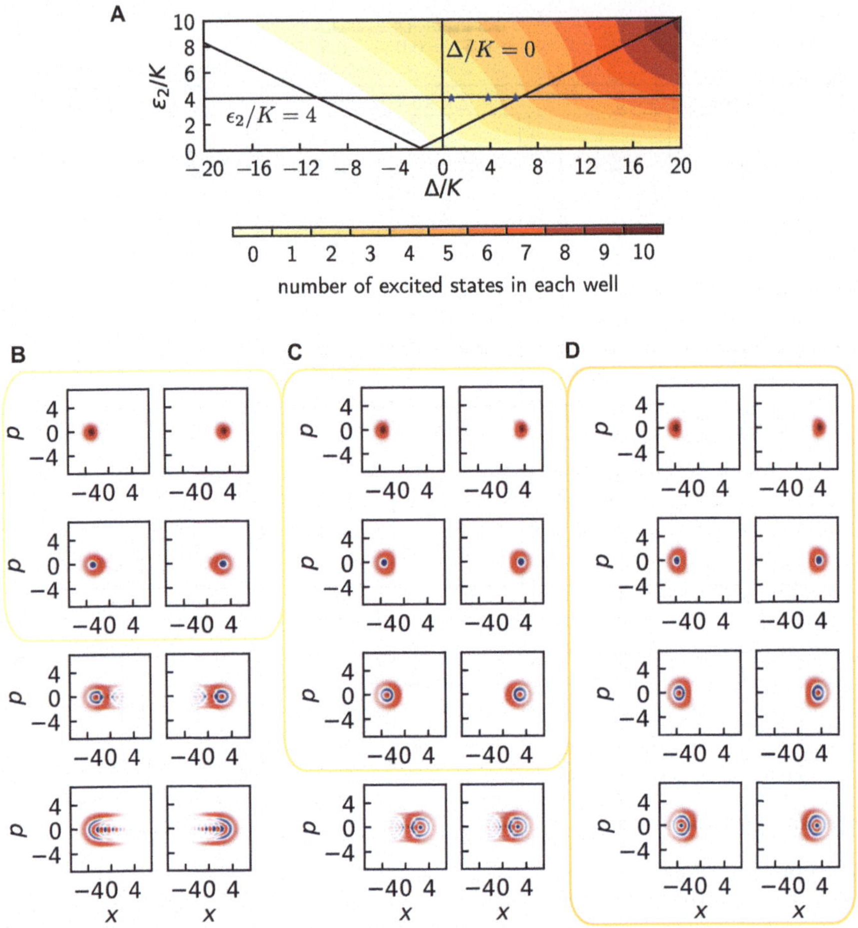

Fig. 3.11 Wigner functions of localized-well states comprising the ground and excited state manifolds of the SKO. (**a**) Semiclassical period-doubling phase diagram Fig. 3.4 with equi-state contours. (**b**)–(**e**) Wigner functions of exact eigenstates' superpositions, corresponding to localized states, for $\epsilon_2/K = 4$, and (**b**) $\Delta/K = 1$, (**c**) $\Delta/K = 4$, (**d**) $\Delta/K = 7$. The action quantization formulation, detailed in Sect. 3.1 and summarized by Eq. (3.10), predicts **B** 1, **D** 2, and **E** 3 excited states respectively. The Wigner functions of states outside this window are seen to have support in the other well too, and larger Δ helps localize them, thus validating the semiclassical picture discussed in Sect. 3.1 quantum mechanically

are fundamentally degenerate and exhibit no tunneling between two wells found in the classical limit (see also [33]). This property can be understood by writing Eq. (3.21) into the factorized form [126]

$$\hat{H}_{\text{PBB}} = -K(\hat{a}^{\dagger 2} - \epsilon_2/K)(\hat{a}^2 - \epsilon_2/K), \tag{3.22}$$

from which it is evident that the two coherent states $|\pm\alpha\rangle$ with $\alpha = \sqrt{\epsilon_2/K}$, which are the eigenstates of the annihilation operator $\hat{a}$, are also degenerate eigenstates of Eq. (3.22). Since Eq. (3.22) is negative-semidefinite and $\hat{H}_{\text{PBB}}|\pm\alpha\rangle = 0$, these states are the ground states.

Note that the Hamiltonians $\hat{H}_{\text{WM}}$ and $\hat{H}_{\text{PBB}}$ differ only by an operator-valued commutator since $\hat{a}^{\dagger 2}\hat{a}^2 - (\hat{a}^\dagger\hat{a})^2 = \hat{a}^\dagger\hat{a}$. Taking the classical limit loses track of this reordering. Their shared classical limit can be written as

$$\begin{aligned} H_{\text{cl}} &= -Ka^{*2}a^2 + \epsilon_2(a^{*2} + a^2) \\ &= -K\left(\frac{x^2+p^2}{2}\right)^2 + \epsilon_2(x^2 - p^2). \end{aligned} \tag{3.23}$$

By introducing the more general Hamiltonian

$$\hat{H} = \Delta\hat{a}^\dagger\hat{a} - K\hat{a}^{\dagger 2}\hat{a}^2 + \epsilon_2(\hat{a}^{\dagger 2} + \hat{a}^2), \tag{3.24}$$

we identify that $\hat{H}_{\text{PBB}}$ and $\hat{H}_{\text{WM}}$ are specific instances of Eq. (3.24) with $\hat{H}_{\text{PBB}} = \hat{H}|_{\Delta=0}$ and $\hat{H}_{\text{WM}} = \hat{H}|_{\Delta=-K}$. Note that taking $\Delta \neq 0$ breaks the simple factorization condition of Eq. (3.22). Indeed, the presence of the $\hat{a}^\dagger\hat{a}$ term is the cause of ground state tunneling in $\hat{H}_{\text{WM}}$, and its absence is the cause of the complete coherent cancellation of tunneling in $\hat{H}_{\text{PBB}}$.

In 2007, Marthaler and Dykman [103] treated a Hamiltonian analogous to Eq. (3.17), and varied a parameter analogous to Δ for a fixed ϵ_2. This led to their prediction of periodic cancellation of tunneling amplitude for the ground state manifold as a function of Δ.

In conclusion, we have introduced in this chapter classical, semiclassical, and quantum analyses to understand properties of our quantum double-well system, the squeeze-driven Kerr oscillator. We uncovered the exponential reduction of tunneling as a function of both ϵ_2 and Δ and multi-level robust spectral degeneracies as a function of Δ. A physical picture to understand the spectral degeneracies in the ground state manifold—as the interference between oscillating wave functions in the classically forbidden region—was developed by Marthaler and Dykman [102, 103]. We are now led to four questions: (1) Can the WKB treatment developed in [102, 103] be extended the excited state manifold? (2) What are the hidden symmetries, if any, that explain the spectral degeneracies as a function of Δ/K? (3) Can such an effective Hamiltonian be engineered in the lab? (4) Can its properties, such as the *spectral kissing* as a function ϵ_2/K and *multi-level degeneracies* as a function of Δ/K be observed? While the first two remain open questions, answering the last two questions will be the focus of the rest of this dissertation.

In the next chapter, we introduce and describe in detail our experimental setup.

Chapter 4
Experimental Setup

Abstract In this chapter, we present the superconducting package and quantum circuit where we have engineered a quantum double-well model system of the squeezed Kerr oscillator (SKO).

First, we describe the package design, then we describe readout in our quantum double-well, and finally the calibration of parameters constituting the effective Hamiltonian Eq. (1.9). In Fig. 4.1, we present a schematic of the superconducting package design. The overarching design goal was to incorporate two capacitively coupled SNAIL-transmon circuits that could each be individually stabilized with a squeezing drive and independently read out. The package consists of a half-aluminum, half-copper package containing two sapphire chips, each with a SNAIL-transmon, readout resonator and Purcell filter. For the work presented here, a summary of the results of [53, 154], we address only one of the two chips with microwave drives and the modes from the other chip play the role of spectators.

Table 4.1 lists a few key design parameters.

Each sapphire chip is clamped to two copper posts with beryllium copper clips, and hosts three electromagnetic modes of interest: SNAIL-transmon, readout resonator, and Purcell filter. A beryllium copper pin inserted into the cavity defines the readout port and sets the linewidth of the readout resonator and Purcell filter; a second weakly coupled pin serves for the application of all microwave drives.

Focusing on the design of the addressed SNAIL-transmon, the design change compared to previous work [66], where the coherent state lifetime saturated at $|\alpha|^2 = 2.6$, is a twenty-fold reduction in self-Kerr nonlinearity. We actuated this reduction by moving to a series array of $M = 2$ SNAILs, reducing the SNAIL junction asymmetry parameter [50], and working at a larger magnetic flux bias $\varphi_{\text{ext}} = 0.33$ in Eq. (2.2), closer to but not at the Kerr-free flux point [51]. These changes resulted in coherent state lifetime saturation around $|\alpha|^2 = \epsilon_2/K = 10$ in the current device (for $\Delta = 0$) as we show in Fig. 6.2.

J. Venkatraman, *Controlling the Effective Hamiltonian of a Driven Quantum Superconducting Circuit*, Springer Theses,
https://doi.org/10.1007/978-3-031-83270-3_4

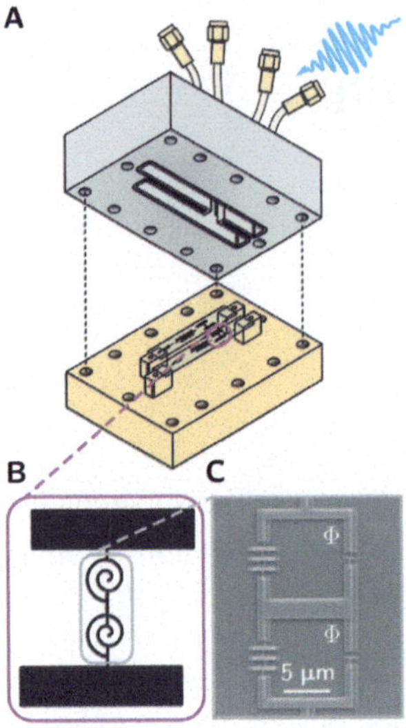

Fig. 4.1 Schematic of the experimental superconducting package. (**a**) Rendering of the half-aluminum, half-copper sample package containing two sapphire chips, each with a SNAIL-transmon, readout resonator and Purcell filter. Only one chip is used in the experiments reported here, first presented in [53, 154]. Applying a strong microwave drive at $\omega_d/2 \approx 2\omega_a$ yields an effective Hamiltonian system of the SKO out of the rapidly driven SNAIL superconducting circuit. (**b**) Schematic of the SNAIL-transmon superconducting circuit: the array of two SNAILs serves as the nonlinear element. (**c**) Scanning electron micrograph (SEM) of the two-SNAIL array. The SNAIL loops are biased with an external magnetic flux $\Phi_{\text{ext}}/\Phi_0 = 0.33$, where Φ_0 is the magnetic flux quantum

4.1 Calibrating the Squeezing Drive Amplitude ϵ_2

In this section, we present a measurement that provides an independent calibration of the squeezing drive amplitude ϵ_2. The pulse sequence is the following: We turn on the squeezing drive at $\Delta = 0$, for a variable amount of time t during which we also turn on a Rabi drive at amplitude ϵ_x and frequency $\omega_d/2 = \omega_a$. The squeezing drive stabilizes the Schrödinger cat states with well-defined parity, and the Rabi drive induces an oscillation in this cat-qubit. We perform this experiment for different values of ϵ_2 and measure $\hat{X} = |\mathcal{C}^+\rangle\langle\mathcal{C}^-| + |\mathcal{C}^-\rangle\langle\mathcal{C}^+|$, where $|\mathcal{C}^\pm\rangle$ are the Schrödinger cat states. This protocol was introduced in [53, 66] and we refer the reader to these works for further details. The result of our experiment is shown in Fig. 4.2a. From this experimental data, we extract a Rabi oscillation frequency Ω_x that is related to the amplitude of the Rabi drive as $\epsilon_x = \Omega_x(\epsilon_2 = 0)/2$. The photon-number at $\Delta = 0$ $|\alpha|_0^2$ is related to ϵ_x and Ω_x as $|\alpha|_0^2 = \Omega_x^2/16\epsilon_x^2$ [53, 66]. In Fig. 4.2b, we plot the experimental data and fit for the extracted photon-number as a function of the digital control amplitude (DAC). With this result, we have a calibration of ϵ_2 as a function of the digital control amplitude (DAC) controlling the squeezing drive.

Table 4.1 Summary of parameters of experimental device. All design simulations (sims.) were performed with Ansys HFSS and black box quantization [112, 117] including corrections to Kerr from cubic nonlinearities [49], which follow from similar corrections in lumped-element calculations [51]. The experiments presented here were performed in the particular flux bias point $\varphi_{ext} = 0.33$. The package details are reproduced here from [49, 53] for the sake of completeness

Parameter	Value	Method of estimate
Oscillator dipole capacitance E_C/h	60 MHz	Design simulation
Oscillator number of SNAILs	2	Design
Oscillator SNAIL asymmetry α	0.1	Resistance msmt.
Oscillator of 1 large junction	0.6 nH	Resistance msmt.
Oscillator inductance of 1 small junction	6 nH	Resistance msmt.
Oscillator frequency at $\Phi/\Phi_0 = 0$	6.668 GHz	Two-tone spec.
Oscillator frequency at $\Phi/\Phi_0 = 0.5$	5.815 GHz	Two-tone spec.
External flux bias point Φ/Φ_0	0.33	Two-tone spec.
Oscillator frequency $\omega_a/2\pi$	6.079 GHz	Two-tone spec.
Oscillator cubic nonlinearity $g_3/3/2\pi$	$\approx$10 MHz	Design sim.
Oscillator self-Kerr nonlinearity $K/2\pi$	316.8 kHz	Two-tone spec. (Fig. 4.3)
Oscillator single-photon decay time T_1	20 μs	Std. coherence msmt.
Oscillator Ramsey decay time T_{2R}	3.86 μs	Std. Ramsey coherence msmt.
Oscillator Hahn echo decay time T_{2E}	13 μs	Std. echo coherence msmt.
Readout resonator frequency $\omega_r/2\pi$	8.506 GHz	Direct RF reflection msmt.
Readout resonator linewidth $\kappa_r/2\pi$	0.40 MHz	Direct RF reflection msmt.
Readout resonator internal linewidth	<0.04 MHz	Direct RF reflection msmt.
Readout to oscillator cross-Kerr $\chi_{ab}/2\pi$	~10 kHz	Design sim.
Purcell filter frequency	8.703 GHz	Direct RF reflection msmt.
Purcell filter linewidth	25 MHz	Direct RF reflection msmt.
Lowest box mode frequency	12.46 GHz	Design sim.

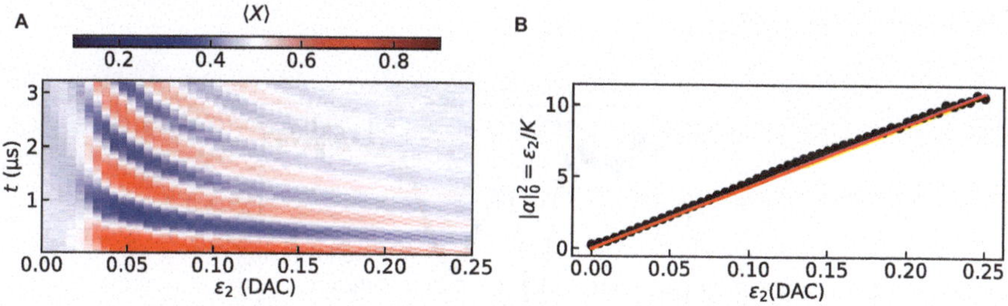

Fig. 4.2 Calibrating ϵ_2 with cat-Rabi oscillations. (**a**) Color plot of $\langle X \rangle$ as a function of the digital control amplitude (DAC) controlling the squeezing drive ϵ_2 and duration of the Rabi drive. We find $\epsilon_x/2\pi = 144.93$ kHz using the relation between the Rabi amplitude and Rabi frequency for $\epsilon_2 = 0$, $\epsilon_x = \Omega_x(\epsilon_2 = 0)/2$. A plot of $|\alpha|_0^2 = \epsilon_2/K = \Omega_x^2/16\epsilon_x^2$ [53, 66] as a function of ϵ_2 in DAC units. A line fit gives us a calibration of $|\alpha|_0^2 = \epsilon_2/K$ as a function of the digital control amplitude (DAC) controlling the squeezing drive

4.2 Measuring the Kerr Coefficient K

In this section, we detail a measurement of the Kerr coefficient K via saturation spectroscopy of the SNAIL transmon. This measurement is performed in the absence of the squeezing drive. In the following text, the letters g, e, and f index the ground, first excited, and second-excited states of the SNAIL transmon oscillator. In Fig. 4.3, we plot the response of the readout as a function of a probe tone, whose frequency is ω_{pr}, and which we vary around the ge transition frequency of the SNAIL transmon oscillator ω_a corresponding to $\epsilon_2 = 0$. When the probe tone excites the oscillator, the readout signal due to the dispersive coupling [10] changes. The two dips in Fig. 4.3, from left to right, correspond to a two-photon transition that excites the oscillator from g to f and to a resonant excitation of the oscillator from g to e respectively. The $gf/2$ and ge resonances are located at $(\omega_a - K)/2\pi$ and $\omega_a/2\pi$ respectively. Fitting the peaks and subtracting their locations yields a value of $K/2\pi = (329.73 \pm 4.30)$ kHz. This value is consistent with the value of $K/2\pi = 316.83$ kHz, where the latter is extracted from Fig. 5.1e in the main text and is the value for K used throughout the article.

Keeping in mind that gate speeds are limited by the gap in the excited state spectrum $4K|\alpha|^2$ and that $K \propto 1/M^2$, gate speeds are independent of the Kerr nonlinearity if the expected $|\alpha|^2$ increase is simultaneously achieved. The exception to this is the Kerr-refocusing gate, which takes time $\pi/2K$ irrespective of $|\alpha|^2$. Generally, the gate fidelity however will decrease under this optimization since the cat-state lifetime $T_{YZ} = 1/2|\alpha|^2 T_1$. As such, we expect there to be an optimum Kerr nonlinearity for a given application depending on the tradeoff between achievable gate speed and noise bias. Improvements in gate design and control techniques will help increase gate speed for a given gap in the excited state spectrum [170], and thereby will allow designs with less nonlinearity to further increase coherent state lifetime and noise bias.

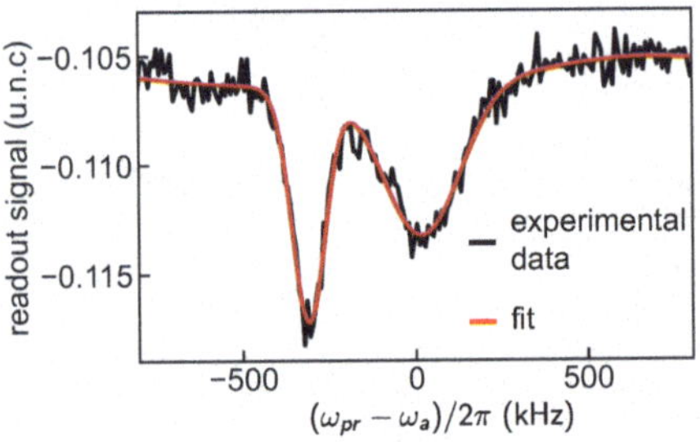

Fig. 4.3 Measurement of K with two-tone saturation spectroscopy. Readout response as a function of the frequency of the saturation (probe) tone. The two readout signal dips in black correspond, from left to right, to the $gf/2$ transition, which is expected to occur at $(\omega_a - K)/2\pi$ and to the ge transition of the SNAIL transmon, which is expected to occur at $(\omega_a)/2\pi$. Here, $gf/2$ refers to a transition induced by two photons from the probe. By fitting the experimental data, we find $K/2\pi = (329.73 \pm 4.30)$ kHz

4.3 Which-Well Readout of the SKO

Measurements of the oscillator state were performed through a separate on-chip readout resonator with frequency $\omega_r/2\pi = 8.5\,\text{GHz}$ and coupling rate $\kappa_r/2\pi = 0.40\,\text{MHz}$ to the quantum-limited measurement chain (see Table 4.1 for device details). In order to activate a frequency-converting beam splitter interaction between the SKO and the readout resonator, we apply an additional drive at $\omega_{BS} = \omega_r - \omega_d/2$. This transfers photons from the SKO to the readout resonator, which are subsequently collected by the measurement chain. The strong squeezing drive ($\epsilon_2 > K$) replenishes these radiated photons, thereby maintaining a steady oscillation amplitude. This is a necessary condition for a quantum nondemolition (QND) measurement of the *which-well* observable $\hat{X} \approx |+\alpha\rangle\langle+\alpha| - |-\alpha\rangle\langle-\alpha|$ [66]. In effect, the readout resonator state is displaced, conditioned on which of the two metapotential wells is occupied.

It is critical for our implementation that we do not rely on the ordinary dispersive readout of the superconducting circuit platform. This means that our Kerr-cats are prepared, evolved, and detected without relying on standard Fock qubit operations or measurements. This ensures the high performance of a system with a weak *bare* nonlinearity. Instead of using the dispersive coupling of the SNAIL-transmon to the readout resonator we use a parametrically activated readout scheme with a large on-off ratio. The readout is enacted by playing a microwave pulse at the frequency difference in between the squeezed Kerr oscillator frame $\omega_d/2$ and readout resonator at ω_r while the stabilization drive is on. The nonlinear term providing the interaction originates from the SNAIL array that, when activated by a microwave tone of displacement amplitude ξ_{BS} and frequency $\omega_{BS} = \omega_r - \omega_d/2$, transforms as

$$g_3\left(\hat{a}+\frac{g_{ba}}{\Delta_{ba}}\hat{b}+\text{h.c.}\right)^3 \rightarrow g_3\left(\hat{a}e^{-i\omega_d t/2}+\xi_{BS}e^{-i\omega_{BS}t}+\frac{g_{ba}}{\Delta_{ba}}\hat{b}e^{-i\omega_r t}+\text{h.c.}\right)^3$$
$$\approx 6g_3\frac{g_{ba}}{\Delta_{ba}}\left(\xi_{BS}\hat{a}\hat{b}^\dagger+\xi_{BS}^*\hat{a}^\dagger\hat{b}\right). \tag{4.1}$$

Here, we have used the RWA to get rid of rapidly rotating terms in a displaced and rotating frame for $\hat{a}$, g_3 is the third order nonlinearity of the SNAILs, g_{ba} is the bare capacitive coupling in between the Kerr-oscillator and the readout resonator, $\hat{b}$ is the annihilation operator of the resonator, and $\Delta_{ba} = \omega_r - \omega_a$. The hybridization is given in the dispersive approximation for simplicity. Note, however, that the RWA approximation is unjustified—the detuning of the readout drive is of order the oscillator frequencies—and higher orders need to be considered. To leading order, their effect is a renormalization of the beamsplitter coupling rate g_{BS} in the effective interaction

$$g_{BS}\hat{a}\hat{b}^\dagger + g_{BS}^*\hat{a}^\dagger\hat{b}, \tag{4.2}$$

which is the announced frequency-converting beamsplitter interaction.

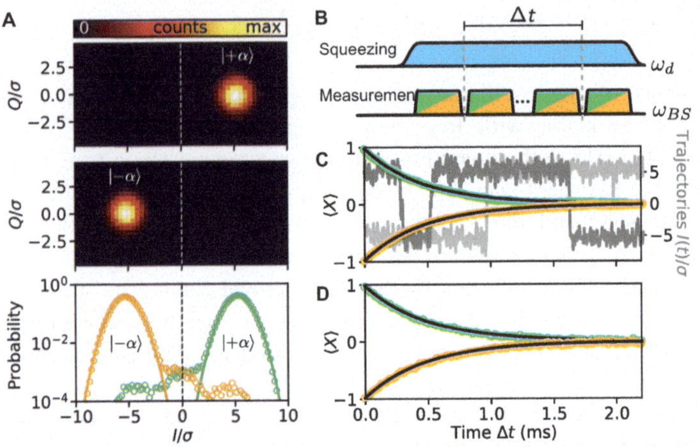

Fig. 4.4 High-fidelity and quantum non demolition (QND) which-well readout of the SKO. (**a**) Top and middle: histogram of the readout resonator output field while performing 2.5×10^8 measurements after preparation in $|\pm\alpha\rangle$ with a previous, stringently thresholded measurement with a bias of 6.5 standard deviations (σ). Bottom: corresponding probability distribution along the I quadrature, and Gaussian fits (solid lines) with standard deviation used to scale the axes. Applying a fair (unbiased) threshold represented by the dashed vertical line yields a readout infidelity of 0.46%. (**b**) Pulse sequence to performing repeated measurements, each with a duration of 4.44 μs. (**c**) Example quantum jump trajectories (grey) under repeated measurements for $\epsilon_2/K = 10.7$. Averages of trajectories conditioned on the first measurement of $|\pm\alpha\rangle$ (green/orange) fit with single exponentials (black) with decay time $T_X^{\text{jumps}} = (485 \pm 1)\,\mu$s. (**d**) State lifetime for $|\pm\alpha\rangle$ (green/orange) with no intermediate measurements (free decay). Black lines are single-exponential fits with decay time $T_X = (482 \pm 4)\,\mu$s. Note that the which-well readout characterization is reproduced here from [49, 53] for the sake of completeness

In Fig. 4.4a, we show a histogram of $\hat{X}$ measurements. The single shot readout infidelity is 0.46%. Correlation measurements determined that the QND infidelity in our experiment is 1.5%. These values mean that we can continuously monitor our system and reconstruct the trajectories associated with the quantum jumps of the well occupation.

In Fig. 4.4b, we show the experimental protocol for measuring the quantum trajectories. After the squeezing drive is turned on, a waiting time equal to $5T_1$ is imposed to let the system adopt a steady-state and a series of measurements is then performed. The sequence of their outcomes constitutes a quantum trajectory record. Two examples of quantum trajectories are shown in shades of grey in Fig. 4.4c. The green and orange data points correspond to averages of 5×10^5 trajectories, each conditioned on the initial measurement falling on the positive or negative side of a threshold defined by the demodulated field quadrature $I = 0$. The decay curve is fitted by a single exponential (black), yielding an incoherent environment-induced activation time of $T_{\pm X}^{\text{jumps}} = (485 \pm 1)\,\mu$s. We next compare these measurements to the free decay of the coherent states $|\pm\alpha\rangle$. This is obtained by performing only two measurements spaced by a variable idling time, in absence of continuous monitoring; the results are shown in Fig. 4.4d. The decay is also fitted by an exponential whose lifetime is found to be $T_{\pm X} = (482 \pm 4)$ µs, thus showing that continuous monitoring does not significantly modify the coherent state timescale T_X in the metapotential.

Having discussed the experimental setup and its readout, we are ready to present the experimental demonstration of the coherent nulling of the tunnel splitting in the ground state manifold in Chap. 5.

Chapter 5
Quantum Tunneling Observations in the Ground State Manifold of the SKO

Abstract In this chapter, we experimentally demonstrate the cancellation of tunneling in the ground state manifold via tunnel-driven time-domain Rabi oscillation measurements. We compare the experimental observations with theory presented in previous chapters to understand the phase-space dynamics of our driven superconducting circuit.

5.1 Discrete Cancellation as a Function of Δ/K

In Fig. 5.1a, we show the classical limit of the metapotential surface for $\Delta/K = 3, \epsilon_2/K = 0.11$, as a function of phase-space coordinates. The arrows indicate the two WKB tunneling paths [102]. Furthermore, we show in Fig. 5.1b, the wavefunctions corresponding to the ground state manifold. Note that these are not the energy eigenstates but their even and odd superpositions, which are localized in the left and right wells. Importantly, in the classically forbidden region, marked in grey, oscillations accompany the expected decay of the wavefunctions [103]. To observe coherent cancellation of tunneling in the ground state manifold, we prepare a localized well state and measure its tunneling probability as a function of time for different values of Δ and ϵ_2. We present the measurement protocol in Fig. 5.1c. The preparation is done by rapidly turning on the squeezing drive until an amplitude of $\epsilon_2/K = 8.7$ is reached. We subsequently wait for $5T_1$ for the system to relax to its steady state in the presence of the squeezing drive and measure, in a quantum nondemolition (QND) manner, the quadrature containing the which-well information. This measurement projects the system into one of the wells. It is done by the microwave activation of a parametric beam splitter interaction between the squeeze-driven Kerr oscillator and a readout resonator strongly coupled to a quantum-limited amplifier chain. We refer the reader to [53] for experimental details, where the preparation-by-measurement procedure for our system was introduced. This readout protocol yields a stabilized fluorescence signal revealing the quadrature measurement outcome, while the squeezing drive sustains the circuit oscillation. After the preparation, we adiabatically lower the squeezing

J. Venkatraman, *Controlling the Effective Hamiltonian of a Driven Quantum Superconducting Circuit*, Springer Theses,
https://doi.org/10.1007/978-3-031-83270-3_5

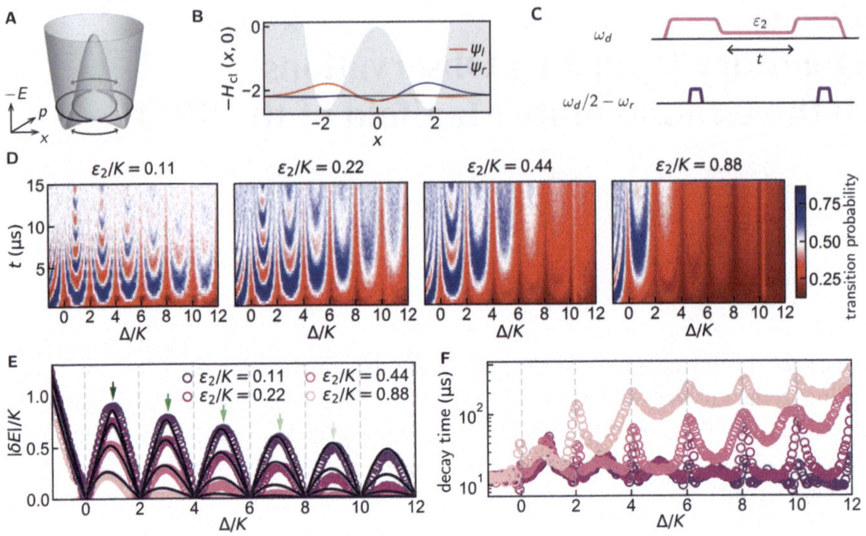

Fig. 5.1 Tunnel-driven Rabi oscillations in the ground state manifold and their periodic nulling. (**a**) Metapotential surface in the classical limit for $\Delta/K = 3$ and $\epsilon_2/K = 0.11$. The orbits shown with black lines are obtained by semiclassical action quantization and represent the ground states (see Chap. 3). Bidirectional arrows represent the two interfering WKB tunneling paths. (**b**) Cut of the classical metapotential surface in (**a**) at $p = 0$. The classically forbidden region is marked in grey. The left and right localized wavefunctions are indicated in red and blue. (**c**) Pulse sequence for (**d**). The pink line represents the squeezing drive at frequency ω_d and the purple lines represent the preparation and readout drives at frequency $\omega_d/2 - \omega_r$. (**d**) Time-domain Rabi oscillation measurement of inter-well tunneling probability (color) as a function of Δ^{bare}, taken here as Δ (see text), for $\epsilon_2/K = 0.11, 0.22, 0.44$, and 0.88. The extracted tunneling amplitudes from (**d**) are shown as open circles in (**e**). The black lines in (**e**) correspond to the transition energy between the lowest eigenstates obtained from an exact diagonalization of Eq. (3.1). A comparison of the extracted tunneling rate with a semiclassical WKB calculation is presented in the supplement. Green arrows in (**e**) denote the condition for constructive interference of tunneling and correspond to the measurements shown in Fig. 5.2. We extract the value of the Kerr coefficient K from this data and note that it is consistent, within experimental inaccuracies, with an independent saturation spectroscopy measurement of the Fock qubit in the absence of the squeezing drive (c.f. Fig. 4.3). (**f**) Decay time of the tunnel-driven Rabi oscillations for different values of Δ and ϵ_2 in (**d**). Sharp peaks in the decay time are clearly visible for $\Delta/K = 2m$, m being a non-negative integer

drive amplitude in a duration 1.6 μs $\gtrsim \pi/K$.[1] The depth of the wells, which increases with ϵ_2/K (see Sect. 3.1), is then reduced so that the tunnel effect becomes observable. We then wait for a variable amount of time before adiabatically raising

[1] Note that this adiabaticity condition pertains to the gap between the ground and first excited pair of states. We do not need to be adiabatic with respect to the two tunnel split states within the ground state manifold since they have opposite parity and the parity preserving squeezing drive will not couple them.

the squeezing drive amplitude to its initial value. Finally, we measure which well the system has adopted.

The data for this tunneling measurement is shown in Fig. 5.1d, where we interpret the oscillating color pattern as tunnel-driven Rabi oscillations. The periodic cancellation of tunneling at $\Delta/K = 2m$, where m is a non-negative integer, is clearly visible as a divergence of the Rabi period. We extract the tunneling amplitude $|\delta E|$ from our data by fitting the oscillation frequency with an exponentially decaying sinusoid and plot this frequency in Fig. 5.1e, where the data-point color corresponds to the value of ϵ_2 (see Chap. 4 for calibration of ϵ_2). The black lines, obtained from an exact diagonalization of the static effective Hamiltonian Eq. (3.1), correspond to the energy difference between levels in the ground state manifold. The cancellation of tunneling for the ground state manifold in a parametrically modulated oscillator was predicted by [103] where, using a semiclassical WKB method, the authors found that this multi-path interference effect is due to, and accompanied by, oscillations of the wavefunction crossing zero in the classically forbidden region. Here, we find good agreement between our experiment and their WKB prediction. Note that, across the zero of the tunneling amplitude, the bonding and anti-bonding superposition of well states alternate as the ground state. Specifically, for $\Delta/K = 4m + 1$, the ground state is the bonding superposition of well states. In Fig. 5.1f, we further plot the extracted decay time of the tunneling oscillations as a function of Δ, and find sharp peaks when $\Delta/K = 2m$, besides an overall continuous increase of the decay time with Δ and ϵ_2. The peaks at $\Delta/K = 2m$ arise from the degeneracies in the excited state spectrum at this condition and are discussed later in the text.

Importantly, the dynamics of the two-level system in Fig. 5.1d suggest a new type of bosonic encoding of information that we call the Δ-Kerr-cat qubit. The north and south poles of the corresponding Bloch sphere, a generalization of the $\Delta = 0$ one [53, 66, 126], is defined by the cat states formed by the lowest pair of eigenstates of Eq. (3.1). In this picture, a tunnel-Rabi cycle in Fig. 5.1d for a fixed $\Delta/K \neq 2m$ corresponds to a travel along the equator. For $\Delta/K = 2m$, this travel is prohibited. Note that when $\Delta/K = 2m + 1$, the tunneling amplitude is maximum and is first-order insensitive to fluctuations of Δ.

5.2 Continuous Reduction as a Function of ϵ_2/K

From Fig. 5.1e, we also see that, besides the *discrete cancellation* of tunneling at $\Delta/K = 2m$, tunneling in the ground state manifold is overall *continuously reduced* with both Δ and ϵ_2. This reflects the well-known symmetry of the double well, which is broken by tunnel coupling. The approximate symmetry is restored with increasing Δ and ϵ_2 because both parameters explicitly control the barrier height and thus exponentially control the tunneling amplitude $|\delta E|$. Theory predicts that the larger the detuning Δ, the faster the tunneling reduction with the squeezing drive amplitude ϵ_2. We have measured this effect by measuring the tunneling amplitude

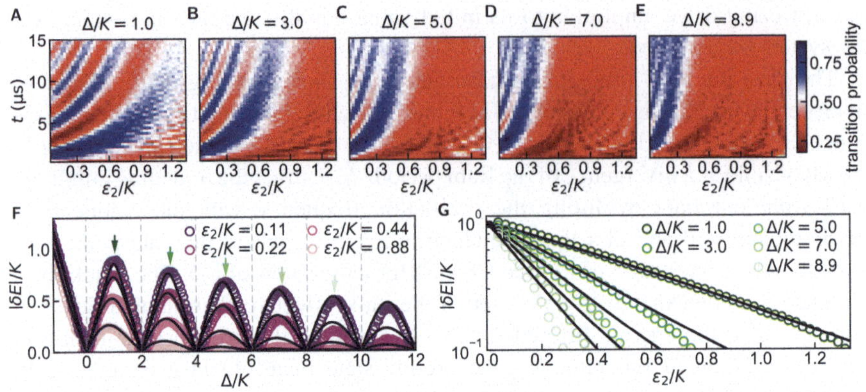

Fig. 5.2 Exponential reduction of tunneling splitting as a function of ϵ_2 in the ground state manifold. For $\Delta/K =$ (**a**) 1, (**b**) 3, (**c**) 5, (**d**) 7, (**e**) 9. (**f**) Extracted tunnel splitting (open circles) for the first five local maxima in Fig. 5.1e as marked by the color coded arrows. Experimental sequence as in Fig. 5.1e. Black lines are obtained from a Hamiltonian diagonalization of Eq. (3.1) with no adjustable parameters. For comparison with a semiclassical WKB calculation, see Fig. 5.3. Note that for small tunneling amplitude, dissipation plays a relevant role and the Hamiltonian model used here is insufficient. (**g**) Extracted tunnel splitting as a function of ϵ_2/K sitting in the hills of (**f**)

as a function of ϵ_2 for different constructive tunneling conditions corresponding to $\Delta/K = 2m + 1$. The data is presented in Fig. 5.2. The exponential insensitivity, around $\Delta = 0$, to fluctuations of Δ due to a noisy ω_a, as a function of ϵ_2, was predicted by [126] and thus proposed as a resource for quantum information. This insensitivity was a key motivation for realizing the Kerr-cat qubit experimentally [66]. The insensitivity of the ground state manifold to detuning as a function of ϵ_2 is directly observed here for the first time. Note from Fig. 5.1e that for $\Delta < 0$, in the parameter regime $\epsilon_2/K < 1$, the tunneling amplitude $|\delta E|$ is weakly dependent on ϵ_2, whereas for $\Delta > 0$, it is strongly dependent on ϵ_2. This weak dependence for $\Delta < 0$ is expected since the barrier height vanishes for small values of ϵ_2/K.[2] Our finding shows that new operating points at even, positive values of Δ/K will increase the resilience of ground-state qubit encoding to detuning-like noise (Fig. 5.3).

In conclusion, in this chapter we presented our experimental observation of the degeneracies in the *ground state manifold* with *time-domain measurements*. Next, in Chap. 6, we will present our experimental observation of multilevel degeneracies in the *excited states*, but with *frequency domain measurements*.

[2] In the absence of dissipation, the metapotential acquires two wells as soon as $\epsilon_2, \ \Delta > 0$, i.e. there is no threshold for bifurcation of the driven oscillator. In our quantum experiment, this threshold is finite but is, relatively speaking, extremely small since and is set by $\epsilon_2^2 > (\Delta^2 + T_1^{-2}/4)/4$ (see Chap. 3).

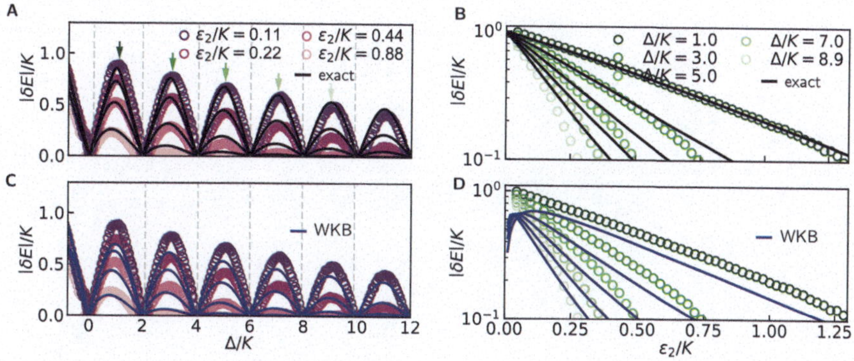

Fig. 5.3 Measurements of tunneling amplitude in the ground state manifold compared to two different models: exact diagonalization of Eq. (3.1) and WKB approximation. The dots in (**a**), (**b**), (**c**), (**d**) is the same data presented in Fig. 5.1e. Solid lines in black in (**a**) and (**b**) are obtained via exact numerical diagonalization. Solid blue lines in (**c**) and (**d**) are obtained via a semi-classical WKB treatment presented in Chap. 3, which is itself adapted from results of Marthaler and Dykman in [102, 103]. Tunnel amplitude in (**a**) and (**c**) are plotted as a function of Δ and in (**b**) and (**d**) as a function of ϵ_2 respectively. As expected, the semi-classical Hamiltonian model, in the domain of its validity $\Delta/K \gg 1$ and $\epsilon_2/K \sim 1$, agrees well with the measured data

Chapter 6
Excited State Manifold: Spectral Kissing, Multilevel Degeneracies, and Their Fingerprint on the Qubit Lifetime

Abstract Moving to the pairs of excited states above the ground state manifold, do they also present observable degeneracies as a function of Δ/K like the ground state manifold? How robust are these degeneracies to perturbations? We present excited state experimental observations and understand them from a quantum and semiclassical perspective.

6.1 Action Quantization in the Excited States

In order to deepen our understanding of the excited state manifolds of our model system, we first examine the classical metapotential surface via the period doubling phase diagram [168] shown in Fig. 3.1. In the classical limit (see Chap. 3), the parameter space spanned by Δ/K and ϵ_2/K is divided by two phase transitions located at $\Delta = \pm 2\epsilon_2$. The different phases are characterized by the number of stable nodes (attractors) in the classical metapotential and we refer to them as the single-, double-, and triple-node phases. These phases correspond to different metapotential topologies. We show them as contour line insets in Fig. 3.1 representing classical orbits. The single-node phase occurs for $\Delta < -2\epsilon_2$, and presents only one well. For $\Delta \geq -2\epsilon_2$, the oscillator has bifurcated and the classical metapotential acquires two wells. In the presence of dissipation, these wells house stable nodes. The emergent ground state manifold has been exploited, for $\Delta = 0$, in the Kerr-cat qubit [53, 66]. In the interval $-2\epsilon_2 \leq \Delta < 2\epsilon_2$, an unstable extremum (saddle point) appears at the origin. For $\Delta \geq 2\epsilon_2$, the saddle point at the origin splits into two saddle points and an attractor reappears at the origin. The barrier height of the classical metapotential is given by $(\Delta + 2\epsilon_2)^2/4K$ in the double-node phase and by $2\epsilon_2\Delta/K$ in the triple-node phase (see Sect. 3.1). To count the number of excited states that have sunk under the barrier, we further introduce in Fig. 3.4 a semi-classical phase diagram of the squeeze-driven Kerr oscillator. Following the Einstein-Brillouin-Keller method, which generalizes the notion of Bohr orbits, we quantize the action enclosed in the metapotential well below the height of the barrier and obtain the number of in-well

J. Venkatraman, *Controlling the Effective Hamiltonian of a Driven Quantum Superconducting Circuit*, Springer Theses,
https://doi.org/10.1007/978-3-031-83270-3_6

excited states. In Fig. 3.4c, we present the corresponding orbits in the metapotential surface for a fixed value of ϵ_2/K, and four values of Δ/K. We validate this simple, semiclassical picture with a fully quantum mechanical calculation of the Wigner functions of localized states in the ground and excited state manifold (see Figs. 3.11 and 3.9). It is clear from this analysis that, by increasing ϵ_2 and Δ, and therefore the barrier height, not only the ground state manifold but even the excited state manifolds become progressively ensconced in the wells, and we thus expect the tunneling between the wells to be drastically reduced.

Spectral Kissing as a Function of ϵ_2/K

To observe the energy levels' dependence on the barrier height [176], we perform spectroscopy of discrete quantum energy levels as a function of the squeezing amplitude. This is achieved by interrupting the idling time, now kept constant, between the readout pulses by a microwave probe tone at frequency ω_{pr}. If ω_{pr} coincides with the energy difference between the ground state and an excited state close to or above the metapotential barrier, an interwell transition becomes likely. In Fig. 6.1a–b we show the measurement of the well transition probability for an initial state localized in one well as a function of squeezing amplitude ϵ_2/K and probe frequency ω_{pr}. In Fig. 6.1c we show the fitted location of the spectroscopic lines as open purple dots. In the same figure, with dashed lines, we also show a numerical diagonalization with no adjustable parameters of the static-effective Hamiltonian Eq. (3.1) with $\Delta = 0$. The agreement between theory and experiment is remarkable given the simplicity of the model. For $|\alpha|^2 \to 0$, we extrapolate the spectrum to that of the bare SNAIL-transmon exhibiting the expected Kerr anharmonic ladder. As the squeezing amplitude—and therefore $|\alpha|^2$—grows, the spectrum becomes pairwise degenerate with levels of different photon-number parity approaching each other in an exponential fashion. We refer to this exponential approach of energy levels as *spectral kissing*. This is, to the best of our knowledge, the first observation of the spectrum of a double-well Hamiltonian as a function of its single control parameter.

To understand the observed spectrum, we employ the semiclassical action quantization argument developed in Sect. 3.1. Following the Bohr-like quantization argument, the area enclosed in units of Planck's constant h counts the a half-integer $N + 1/2$ number of quantum states in the wells. Analytically, we find $N = \epsilon_2/\pi K = |\alpha|^2/\pi$ (see Sect. 3.1). Every time N coincides with an integer value, a new pair of excited semiclassical orbits is captured by the wells. The vertical dashed lines in Fig. 6.1b and Fig. 6.2a,b correspond to this semiclassical condition. As the squeezing amplitude ϵ_2 increases, the captured levels sink under the metapotential barrier and, coupled only by quantum tunneling, exponential level kissing takes place.

Besides the overall *continuous reduction* of tunneling, the excited state manifold of the squeeze-driven Kerr oscillator experiences a *discrete cancellation* of tunneling when $\Delta/K = 2m$. Since the squeezing interaction preserves photon

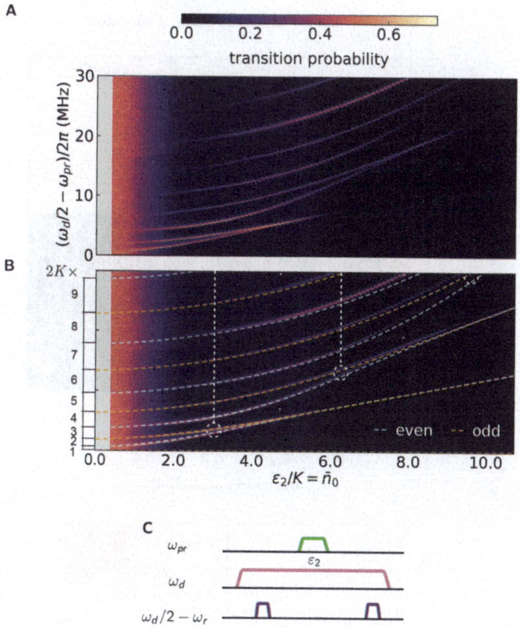

Fig. 6.1 (**a**) Excited state spectroscopy of the SKO Hamiltonian Eq. (3.1) as a function of ϵ_2 with $\Delta = 0$, which models a quantum double-well model system. Color represents the well-flip transition probability. (**b**) Experimental data in (**a**) superposed with theoretical simulation of the tunnel splitting obtained by an exact diagonalization of Eq. (3.1) with no adjustable parameters. Circles and vertical lines mark values of ϵ_2/K where a new pair of excited states become ensconced inside the double-well following a Bohr quantization argument detailed in Sect. 3.1. The vertical axis on the left marks the spacing between consecutive lines for $\epsilon_2 = 0$ given by integer multiples of $2K$. (**c**) Pulse sequence for the measurement presented in (**a**). We note that this figure is an adaptation from [49]

parity, levels belonging to the even and odd sector of the Kerr Hamiltonian remain decoupled and repeatedly cross at values of Δ/K corresponding to even integers. This braiding induces $m + 1$ perfect degeneracies at $\Delta/K = 2m$. Moreover, the corresponding eigenstates have a closed-form expression in the Fock basis. Remarkably, these features are independent of the value of ϵ_2, reflecting a particular, unappreciated symmetry of our Hamiltonian Eq. (3.1) (see Sect. 3.1).

Staircase in T_X as a Function of ϵ_2/K

The most remarkable consequence of the pairwise level kissing is the staircase-shaped increase of the lifetime T_X as a function of the squeezing amplitude ϵ_2/K, as shown in Fig. 6.2. This step-wise increase can be understood, to a first approximation, using Bohr's action quantization. For $\epsilon_2 \ll K$, T_X corresponds

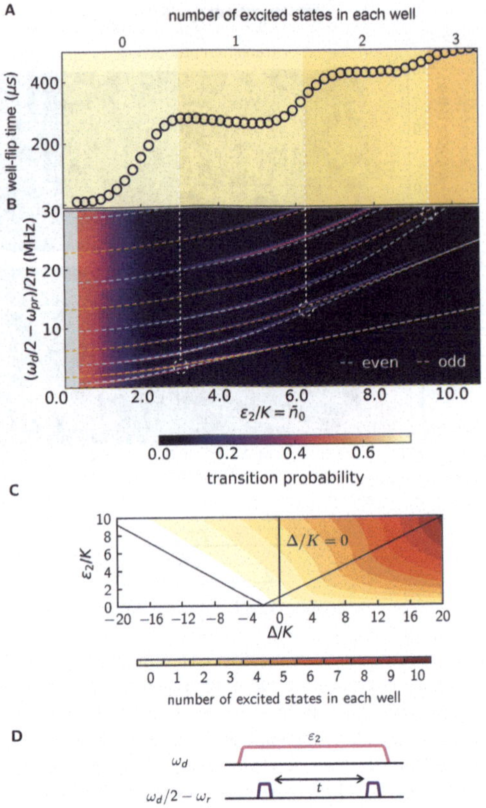

Fig. 6.2 Measured well-flip time of the SKO aka the coherent state lifetime (T_X) of the Kerr-cat qubit as a function of photon-number $\epsilon_2/K = \bar{n}_0$. The value of T_X for each value of $\bar{n}_0$ is obtained by fitting a decaying exponent to the measurement data obtained by following the pulse sequence shown in (**d**). A quantitative model to explain the staircase structure is presented in Chap. 7. Qualitatively, the first step occurs due to the exponentially reducing overlap between the exactly degenerate coherent states. Subsequent steps occur when a pair of excited states become ensconced in the well and further get quasidegenerate. When the excited states become quasidegenerate, thermal-assisted tunneling through excited states get suppressed adding an extra layer of protection to the Kerr-cat qubit. This qualitative argument is confirmed by the coinciding locations of the steps in the staircase in (**a**), the spectral kissing marked as circles in (**b**) and the semiclassical Bohr quantization condition marked as the top X-axis in (**a**). The colors in (**a**) mark the number of excited states in the well following the semiclassical period-doubling phase diagram in (**c**). We note that panels (**a**) and (**b**) have been adapted from [49]

to loss of coherence of the superposition between the ground state and the first excited state of a Kerr oscillator (T_{2R}), but increases exponentially as each of the two metapotential barrier becomes deep enough to host one quantum each ($\epsilon_2/\pi K \sim 1$. The exponential increase stops when excitations to the first pair of excited states ($\hat{a}^\dagger$-like events) become the limiting factor. An excitation into these states will allow the transition between wells. The incoherent environment-induced activation time thus

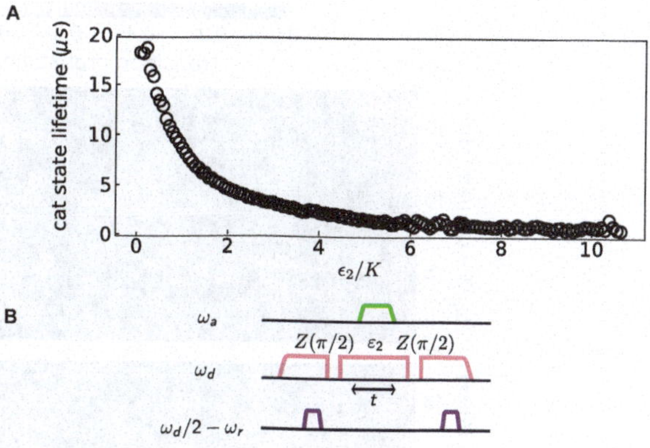

Fig. 6.3 (**a**) Cat state lifetime T_{YZ} of the Kerr-cat qubit as a function of photon-number $\epsilon_2/K = |\alpha|^2$. (**b**) Ramsey-like pulse sequence for the measurement result presented in (**a**). We note that panels (**a**) and (**b**) have been adapted from [49]

plateaus at $\sim$250 μs until Bohr's quantization condition is met again and the first pair of excited states is captured by the metapotential wells ($\epsilon_2/\pi K \sim 2$, second vertical dashed line). At this point, the tunnel splitting between the first pair of excited states vanishes and the increase of lifetime resumes. This cycle repeats itself for the next pair of excited states as shown by the third rising slope in lifetime at $\epsilon_2/\pi K \sim 3$ (third vertical dashed line). We thus interpret the experimental data in Fig. 6.2 as a second manifestation of the pairwise kissing spectrum. In Chap. 7, we go beyond a Hamiltonian reasoning, by providing a Lindblad master equation model that takes into account single-photon heating and loss, which models the experimental data. In Fig. 6.3, we present a measurement of the lifetime T_{YZ} of the Schrödinger cat superpositions of $|\pm\alpha\rangle$, which are all degenerate ground states of Eq. (3.1) for $|\alpha| > 0$. The experimental sequence used for this measurement is shown in Fig. 6.3b. The lifetime T_{YZ} is measured as the decay time of oscillation between cat states.

In this section we presented the spectroscopy of the excited states in the quantum double-well as a function of ϵ_2/K. Note that this corresponds to the experimental validation of Fig. 3.5a. We further presented measurements of the stepped variation T_X as a function of ϵ_2/K, a fingerprint of the excited state spectrum. We will next present spectroscopy measurements as a function of Δ/K for a fixed value of ϵ_2/K corresponding to the experimental validation of Fig. 3.5b.

Multilevel Spectral Degeneracies as a Function of Δ/K

Both the *discrete cancellation* and the overall *continuous reduction* of tunneling now in the excited state manifold of the squeeze-driven Kerr oscillator is accessed by performing spectroscopy measurements as a function of Δ, which we show in Fig. 6.4 for $\epsilon_2/K = 2.17$. The measurement protocol is shown in Fig. 6.1c. We prepare a localized well state in a manner that is similar to the protocols of Figs. 5.1 and 5.2. To locate the frequency of the excited states, we apply a probe

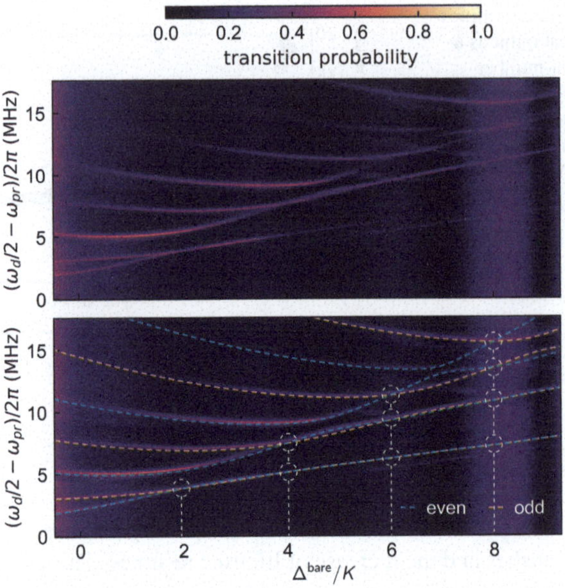

Fig. 6.4 Frequency-domain measurement of well-transition probability (color) via excited states as a function of Δ/K for $\epsilon_2/K = 2.17$. The pulse sequence is shown in Fig. 6.1c. The power of the perturbative spectroscopic probe is increased as ω_{pr} is decreased to compensate for the lower matrix element connecting the ground state with the higher excited levels, yet is kept weak enough to preserve the parity conservation rules of Eq. (3.1). Dashed lines in lower panel are plotted on top of experimental data (same as upper panel) and correspond to transition energies obtained by performing an exact diagonalization of Eq. (3.1) with no adjustable parameters. The Kerr coefficient is calibrated via time-domain measurements in Fig. 5.1e

tone at variable frequency in the vicinity of the SNAIL transmon resonance ω_a and measure the well-switching probability. When the probe is resonant with a transition to a state close to the barrier maximum, this probability is increased. The experimental results are shown in Fig. 6.4. The colored dashed lines (orange and blue) in the lower panel are obtained from an exact diagonalization of the static effective Hamiltonian Eq. (3.1) with no adjustable parameters. The crossings of levels are marked with circles. The data also shows that the level crossings are accompanied by a continuous reduction of the braiding amplitude with Δ. The corresponding reduction of the tunnel splitting is the manifestation associated with a generic double-well Hamiltonian while the braiding reflects interference specific to our particular Hamiltonian, resulting from its underlying driven character. The level of experimental control achieved allows us to observe in this data the joint presence of the *exact discrete symmetry* and the *approximate continuous symmetry* in our bosonic system.

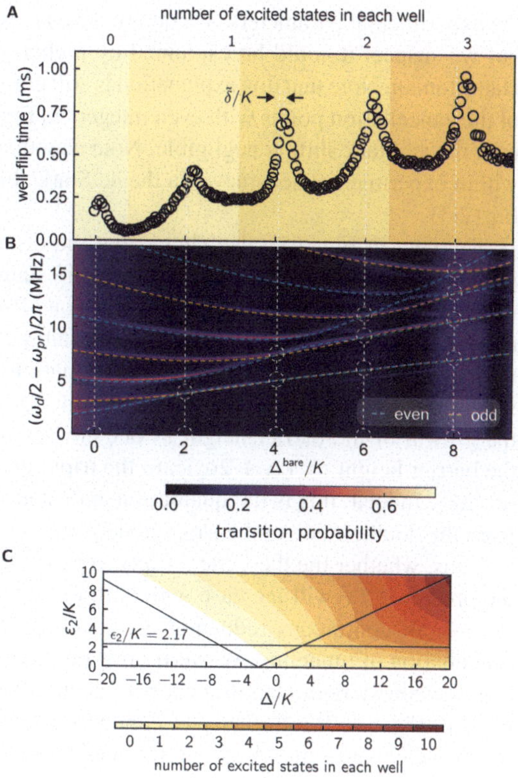

Fig. 6.5 (**a**) Measured well-flip time of the SKO aka the coherent state lifetime (T_X) of the Δ-Kerr-cat qubit as a function of Δ/K for $\epsilon_2/K = 2.17$. At $\Delta/K = 2m+1, m+1$ degeneracies occur in the spectrum as shown in (**b**). Thus, the Kerr-cat qubit is mth order protected to tunneling through excited states, which manifests as increasing peaks in the lifetime. Background color in (**a**) marks the number of excited states per well following semiclassical orbit quantization. A detailed modeling of the decoherence of the SKO is done in Chap. 7. (**c**) Semiclassical period-doubling phase diagram

Peaks in T_X as a Function of Δ/K

An important consequence of the cancellation of tunneling in the excited state spectrum is the periodic enhancement of the well-switching time under incoherent environment-induced evolution. This time scale corresponds to the transverse relaxation time, T_X, of a new bosonic qubit: a Δ-variant of the Kerr-cat qubit [49, 126] as mentioned earlier. To measure T_X, we prepare a localized well state by measurement, and wait for a variable amount of time before measuring the which-well information. We show the pulse sequence in Fig. 6.3b. We obtain T_X by fitting a decaying exponential function to the measured well-transition probability for each value of Δ and plot the result in Fig. 6.5a. Note that we have chosen the squeezing drive amplitude, identical to that of Fig. 6.4a, as $\epsilon_2/K = 2.17$. Around values of Δ/K corresponding to even integers, the variation of T_X presents sharp peaks. The location of the peaks corresponds to the degeneracy condition in the excited state spectrum, associated with coherent cancellation of tunneling and the blocking of noise-induced well-switching pathways via the excited states. The systematic right-offset $\tilde{\delta}/K$ of each peak from an even integer, is 15%. About 5% can be attributed to the ac Stark shift δ^{ac} for this photon number, given the accuracy of our knowledge of the experimental parameters. We do not have an explanation for the remaining 10%, but we suspect it could be explained by higher-order terms in our static effective Hamiltonian. Note that this explanation is still compatible with the perfect alignment of the cancellation points with even integers in Fig. 5.1f for $\epsilon_2/K < 1$, since for that case the ac Stark shift is negligible. Note also that this offset could provide access, within experimental accuracy, via the ac Stark shift, to the nonlinear parameters of Eq. (2.3).

The data in Fig. 6.5a also shows that the discrete peaks are accompanied by a monotonic baseline increase, a direct manifestation of the overall continuous tunneling reduction in the spectrum as a function of Δ. The background colored stripes represent the number of in-well excited states found via the action quantization method discussed above and in Sect. 3.1. Continuing with this semiclassical picture, we interpret the slowdown in the growth of T_X for $\Delta/K \gtrsim 5$ as the slowdown in the growth of the barrier height as one crosses over from the double-node, where the barrier height $\propto (\Delta + 2\epsilon_2)^2$, to the triple-node phase, where the barrier height $\propto \Delta\epsilon_2$. Indeed, this is the quantum manifestation of the classical phase transition from the double-node to the triple-node phase.

Thus, whether the theoretical framework is classical, semiclassical, or quantum, the predicted T_X will increase with both ϵ_2 and Δ. While ϵ_2 and Δ contribute via the overall continuous reduction of tunneling [53], only Δ controls the discrete cancellation of tunneling. We verify this prediction by measuring T_X while varying simultaneously both Hamiltonian parameters. We present the result of this experiment in Fig. 6.6. We further plot contours of constant barrier height in black, and the expected separation between the double-node and triple-node metapotential as

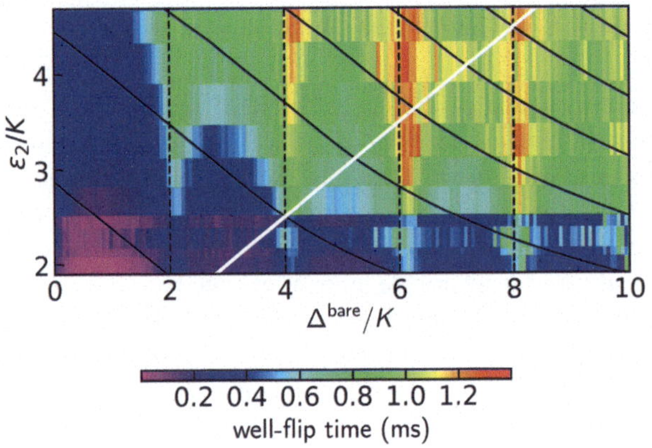

Fig. 6.6 Color map of T_X in the parameter space of Δ/K and ϵ_2/K. White line, given by $\Delta/K + 2 = 2\epsilon_2/K$, marks the transition from a two-node to a three-node metapotential. Black solid lines mark contours of constant barrier height of the metapotential. Increasing both Δ/K and ϵ_2/K yields fastest enhancement in T_X as predicted by the semiclassical phase diagram Fig. 6.5b. The additional enhancement by the coherent cancellation of excited state tunneling at $\Delta/K = 2m$ stands out. The pulse sequence for the measurement is shown in Fig. 6.5c

a white line. The system lying deeply in the quantum regime, we do not expect any sharp features along this line. As expected, following the gradient of the barrier height, one observes the fastest gain in T_X, with a maximum of $T_X = 1.3$ ms for $\Delta/K = 6$ and $\epsilon_2/K = 4$. Increasing the lifetime by increasing ϵ_2 presents limitations, since strong drives are known to cause undesired effects in driven nonlinear systems (see [152, 170] and Chap. 7).

Bistability for non-zero Δ was predicted by Roberts and Clerk in [133]. Our work demonstrates this bistability experimentally through the lifetime peaks in Fig. 6.5a and explains the peaks as a fingerprint of the observed spectral degeneracies in Fig. 6.4. Furthermore, the resilience to noise in the non-zero Δ case is demonstrated through Figs. 5.1e and 5.2.

One could argue that $\Delta = 0$ provides an important factorization condition that guarantees that the ground state manifold is spanned by exact coherent states (see [126] and Fig. 3.10). Indeed, this is an asset for quantum information, since these states are eigenstates of the single-photon loss operator $\hat{a}$ [114]. However, this desirable property is traded for the advantages discussed earlier when $\Delta/K = 2m$, $m \geq 1$. Even if the Δ-variant of the Kerr-cat qubit suffers from quantum heating and quantum diffusion [44, 102, 120] at zero temperature resulting from the squeezed nature of its ground states, these effects are small and, as we show in the experiments reported here and in [53], the well-states of the Kerr-cat live longer than its $\Delta = 0$ parent, even at finite temperature.

In the discussion thus far, we have provided an intuitive picture with which to think about the well-flip lifetime, which corresponds to the coherent-state lifetime of the Δ-variant of the Kerr-cat qubit. Next, we derive a model to quantitatively explain the experimental data. For the sake of concreteness, we will model the data presented in Fig. 6.2a corresponding to the coherent state lifetime of the Kerr-cat qubit ($\Delta = 0$).

Chapter 7
A Decoherence Model for the SKO: An RWA Model and Treating Effects Beyond the RWA

Abstract In this chapter, we capture the effect of the environmental fluctuations on our Kerr parametric oscillator by computing the effective Lindbladian of a Kerr parametric oscillator beyond the rotating wave approximation. We find that the relevant dissipators depend strongly on the nonlinearity and the mean-photon number in the oscillator. In particular, we address recent measurements of the logical lifetime on a Schrödinger cat qubit stabilized by a parametric drive, thereby providing an explanation for the observed three-order-of-magnitude discrepancy between experimental data and an ordinary Lindblad treatment. Our systematic approach paves the path to treat dissipative effects in Floquet-engineered nonlinear quantum systems.

7.1 Introduction

Out-of-equilibrium driven nonlinear oscillators have attracted interest as a platform for tailoring Floquet Hamiltonian spectra and stabilizing quantum states for quantum information. However, it is usually ignored that the necessary parametric drive can amplify environmental fluctuations limiting applications. Static effective Hamiltonians can be engineered in circuit quantum electrodynamics [10] by coherently driving parametric processes. Such techniques have been put to use in creating qubits [96, 97, 126, 156], gates between them [8, 55, 92, 134], readout schemes [47, 90, 150], and quantum simulations [7, 14, 85, 157]. Similar techniques are employed in quantum simulation with atomic systems [58, 104, 166]. Effective Hamiltonians resulting from complex pulse sequences in Trotterization schemes applied to a system [19, 104, 136] can be also viewed as belonging to the above class.

Since physical systems are inevitably open, the nonlinear mixing processes associated with the Hamiltonian parametric terms of interest are also driven incoherently by fluctuations of the environment. These environmental fluctuations can be thermal in origin, in which case the process can be understood as a classical nonlinear mixing of noise that is down- or up-converted to the frequency of the nonlinear

J. Venkatraman, *Controlling the Effective Hamiltonian of a Driven Quantum Superconducting Circuit*, Springer Theses,
https://doi.org/10.1007/978-3-031-83270-3_7

oscillator, or have an origin in the vacuum fluctuations of the environment. These vacuum fluctuations can be amplified by the drive and give rise to heating even in a zero temperature environment, a phenomenon known as Unruh heating when the driving force produces a simple time-independent acceleration [12, 151, 165].

A recent work [121] studied these effects in an attempt to explain drive-induced lifetime reduction in transmon circuits during readout. But in transmons, these effects tend to be masked by multiphoton nonlinear resonances limiting readout and parametric operations [10, 34, 138, 142]. However, the recent implementation of a SKO giving rise to the Kerr-cat qubit [53, 66, 126] provides an ideal platform to uncover the effect of drive-enhanced environmental fluctuations, since unwanted nonlinear resonances of the transmon qubit are largely absent in this new system. Mixing of the environmental fluctuations is captured by beyond rotating wave approximation (RWA) in corrections to the system-bath coupling, giving rise to modified Lindbladian dynamics. In this chapter, we compute the static effective dissipators for the Kerr-cat system and discuss possible new effects that may explain experimental data in [53]. Our systematic method, based on [153], can be extended to arbitrary order and can be applied to other controllable driven systems with a residual coupling to a bath. For the sake of concreteness, we will model the measurements of the coherent state lifetime of the Kerr-cat qubit as a function of ϵ_2/K presented in Fig. 6.2a corresponding to taking $\Delta = 0$.

Decoherence in a Rapidly Driven Nonlinear System

The starting point of the calculation is the driven system-bath Hamiltonian

$$\hat{H}_{\text{tot}}(t) = \hat{H}_{\text{s}} + \hat{H}_{\text{b}} + \hat{H}_{\text{sb}} + \hat{H}_{\text{d}}(t). \tag{7.1}$$

The system is a weakly nonlinear oscillator whose Hamiltonian $\hat{H}_{\text{s}}$ is given by Eq. (2.3). Here, $\hat{a}$ is the bosonic annihilation operator. The parameters ω_o and $g_n \ll \omega_o$ are the bare oscillator frequency and the n-th rank nonlinearity coefficients of the oscillator. We specialize our calculation to the case of the Josephson cosine potential as a source of oscillator nonlinearity and thus take the nonlinear coefficient g_n of the Hamiltonian expansion to be of order $\varphi_{\text{zps}}^{n-2}$ [153], where φ_{zps} is the zero point spread of the phase across the Josephson junction $\hat{\varphi} = \varphi_{\text{zps}}(\hat{a} + \hat{a}^\dagger)$. The system is driven by $\hat{H}_{\text{d}}(t) = -i\hbar F(t)\left(\hat{a} - \hat{a}^\dagger\right)$, where $F(t)$ is the waveform of the drive. At this time, we limit our analysis to the modeling of experiments in which the time dependence of the Hamiltonian corresponds to a monochromatic drive $F(t) = \Omega_d \cos(\omega_d t)$. The environment is taken to be a bath of linear oscillators with Hamiltonian $\hat{H}_{\text{b}} = \sum_j \hbar\omega_j \hat{b}_j^\dagger \hat{b}_j$, which couples to the system by $\hat{H}_{\text{sb}} = -\left(\hat{a} - \hat{a}^\dagger\right)\sum_j h_j\left(\hat{b}_j - \hat{b}_j^\dagger\right)$. In these expressions $\hat{b}_j$ is the annihilation operator of a bath mode at frequency ω_j.

Effective Lindbladian at Order φ^0_{zps}: The RWA Lindbladian

Motivated by the SKO (see Chap. 2,[25, 53, 66, 126, 127, 133, 161]) and quantum information processing with cat-qubits [25, 56, 66, 93, 96, 97, 126, 127, 129], we look now for the static effective description of $\hat{H}_{\text{tot}}$ under the condition $\omega_d \approx 2\omega_o$. The construction of this effective description involves successive unitary transformations followed by averaging out the rapidly oscillating terms in the new frame. First, following [153], we rewrite $\hat{H}_{\text{tot}}$ in a new frame comprising (i) a displaced frame relative to the linear resonance of the oscillator to the drive so that $\hat{a} \to \hat{a} + \frac{i\Omega_d}{2(\omega_d-\omega_o)}e^{-i\omega_d t} - \frac{i\Omega_d}{2(\omega_d+\omega_o)}e^{i\omega_d t}$, (ii) a rotating frame of $\hat{a}$ mode at $\omega_d/2$ so that $\hat{a} \to \hat{a}e^{-i\omega_d t/2}$, and then (iii) a rotating frame of each $\hat{b}_j$ mode at frequency ω_j so that $\hat{b}_j \to \hat{b}_j e^{-i\omega_j t}$. The Hamiltonian now reads

$$\hat{H}_{\text{tot}} = \hat{H}_{\text{s}}(t) + \hat{H}_{\text{sb}}(t) \tag{7.2a}$$

$$\begin{aligned}\hat{H}_{\text{s}}(t)/\hbar = \delta\hat{a}^\dagger\hat{a} + \sum_n \frac{g_n}{n}\Big(&\hat{a}e^{-i\omega_d t/2} + \hat{a}^\dagger e^{i\omega_d t/2} \\ &+ \Pi e^{-i\omega_d t} + \Pi^* e^{i\omega_d t}\Big)^n\end{aligned} \tag{7.2b}$$

$$\hat{H}_{\text{sb}}(t) = i\left(\hat{a}e^{-i\omega_d t/2} - \hat{a}^\dagger e^{i\omega_d t/2}\right)\hat{B}(t), \tag{7.2c}$$

where $\delta = \omega_o - \omega_d/2$, $\Pi \approx 4i\Omega_d/3\omega_d$, and $\hat{B}(t) = \sum_j i h_{\omega_j}\left(\hat{b}_j e^{-i\omega_j t} - \hat{b}_j^\dagger e^{i\omega_j t}\right)$. Averaging out the fast oscillation arising in $\hat{H}_{\text{s}}(t)$, one finds the system Hamiltonian and its coupling to the environment under the RWA (order φ^0_{zps}). We further replace the sum $\sum_j$ over the bath modes with an integral introducing a density of modes λ_ω such that $\lambda_\omega d\omega$ gives the number of oscillators with frequencies in the interval from ω to $\omega + d\omega$. Tracing out the environment at this point under the usual Born-Markov approximation in a thermal bath provides the ordinary Lindbladian [25, 56, 129], which involves the usual dissipators corresponding to single photon loss $\mathcal{D}[\hat{a}]$ and gain $\mathcal{D}[\hat{a}^\dagger]$ [16, 21], where $\mathcal{D}[\hat{O}]\bullet := \hat{O}\bullet\hat{O}^\dagger - (\hat{O}^\dagger\hat{O}\bullet + \bullet\hat{O}^\dagger\hat{O})/2$. The effect of the bath under the Markov approximation is equivalent to a stochastic force coupled to the system by $i\hat{f}(t)(\hat{a} - \hat{a}^\dagger)$ with spectral density $S_{ff}[\omega] = 2\pi\lambda_\omega|h_\omega|^2\bar{n}_\omega$, $S_{ff}[-\omega] = 2\pi\lambda_\omega|h_\omega|^2(1+\bar{n}_\omega)$, where $\bar{n}_\omega$ is the average photon number of the mode $\hat{b}_\omega$ at frequency $\omega > 0$ [32].

The key to obtaining our main result is to take into account terms beyond the RWA in the system-bath coupling and get an averaged description of $\hat{H}_{\text{tot}}$. We follow our generalization of the Schrieffer-Wolff transformation procedure [153] to construct a near-identity canonical transformation generated by $\hat{S}(t) = \mathcal{O}(\varphi_{\text{zps}})$ so that the transformed Hamiltonian is time-independent to order φ^k_{zps} for some arbitrarily large k of interest. Under $\hat{S}$, $\hat{H}_{\text{tot}}(t) \to \hat{\mathcal{H}}_{\text{eff}}$, which is given as

$$\begin{aligned}\hat{\mathcal{H}}_{\text{eff}} &\equiv e^{\hat{S}/i\hbar}\hat{H}_{\text{tot}}(t)e^{-\hat{S}/i\hbar} - i\hbar e^{\hat{S}/i\hbar}\partial_t e^{-\hat{S}/i\hbar} \\ &= \hat{\mathcal{H}}_{\text{s}} + \hat{\mathcal{H}}_{\text{sb}},\end{aligned} \tag{7.3}$$

where, by construction [153], $\hat{\mathcal{H}}_{\text{eff}}$ is the static effective approximation of $\hat{H}_{\text{tot}}(t)$, and the computation of $\hat{S}(t)$ is detailed in Sect. 7.1. The first summand in Eq. (7.3) reads

$$\hat{\mathcal{H}}_{\text{s}}/\hbar = \Delta\hat{a}^{\dagger}\hat{a} - K\hat{a}^{\dagger 2}\hat{a}^{2} + \epsilon_2(\hat{a}^{\dagger 2} + \hat{a}^{2}) + \mathcal{O}(\varphi_{\text{zps}}^{3}), \tag{7.4}$$

where $\Delta = \delta + 6g_4|\Pi|^2 - 18g_3^2|\Pi|^2/\omega_d + 2K$ is the Stark- and Lamb-shifted detuning, $K = -3g_4/2 + 20g_3^2/3\omega_d$ is the Kerr coefficient, and $\epsilon_2 = g_3\Pi$ is the squeezing amplitude.

Effective Lindbladian at Order φ_{zps}^1: First Order Beyond the RWA

The canonical transformation generated by $\hat{S}(t)$ can be viewed as describing the system in an accelerated frame. In this frame, the system effectively experiences the static Hamiltonian Eq. (7.4); meanwhile, the system-bath coupling develops nonlinear components. Keeping terms to order φ_{zps}^1, the perturbation parameter in the expansion of $\hat{S}(t)$, the system-environment coupling reads

$$\begin{aligned}\hat{\mathcal{H}}_{\text{sb}}^{(1)} \approx\ & i\left(\hat{a}e^{-i\omega_d t/2} - \hat{a}^{\dagger}e^{i\omega_d t/2}\right)\hat{B}(t) \\ & + i\Big(-\frac{3\epsilon_2}{\omega_d}\hat{a}^{\dagger}e^{i3\omega_d t/2} - \frac{8g_3}{3\omega_d}\hat{a}^{\dagger 2}e^{i2\omega_d t/2} \\ & - \frac{2\epsilon_2}{\omega_d}\hat{a}e^{i\omega_d t/2} + \frac{2\epsilon_2}{\omega_d}\hat{a}^{\dagger}e^{-i\omega_d t/2} \\ & + \frac{8g_3}{\omega_d}\hat{a}^{2}e^{-i2\omega_d t/2} + \frac{3\epsilon_2}{\omega_d}\hat{a}e^{-i3\omega_d t/2}\Big)\hat{B}(t),\end{aligned} \tag{7.5}$$

where the first line, at order φ_{zps}^0, is identical to the coupling term Eq. (7.2c).

Following a standard Lindbladian derivation [16, 21, 22], but now with the renormalized system-bath Hamiltonian, we obtain the effective Lindblad master equation for the system up to order φ_{zps}^1 as

$$
\begin{aligned}
\partial_t \hat{\rho}_s =& \frac{1}{i\hbar}[\hat{\mathcal{H}}_s, \hat{\rho}_s] + \kappa_{\omega_d/2}\bar{n}_{\omega_d/2}\mathcal{D}[\hat{a}^\dagger + \frac{2\epsilon_2}{\omega_d}\hat{a}]\hat{\rho}_s \\
&+ \kappa_{\omega_d/2}\left(1+\bar{n}_{\omega_d/2}\right)\mathcal{D}[\hat{a} + \frac{2\epsilon_2}{\omega_d}\hat{a}^\dagger]\hat{\rho}_s \\
&+ \kappa_{\omega_d}\bar{n}_{\omega_d}\left(\frac{8g_3}{3\omega_d}\right)^2 \mathcal{D}\left[\hat{a}^{\dagger 2}\right]\hat{\rho}_s \\
&+ \kappa_{\omega_d}\left(1+\bar{n}_{\omega_d}\right)\left(\frac{8g_3}{3\omega_d}\right)^2 \mathcal{D}\left[\hat{a}^2\right]\hat{\rho}_s \\
&+ \kappa_{3\omega_d/2}\bar{n}_{3\omega_d/2}\left(\frac{3\epsilon_2}{\omega_d}\right)^2 \mathcal{D}\left[\hat{a}^\dagger\right]\hat{\rho}_s \\
&+ \kappa_{3\omega_d/2}\left(1+\bar{n}_{3\omega_d/2}\right)\left(\frac{3\epsilon_2}{\omega_d}\right)^2 \mathcal{D}[\hat{a}]\hat{\rho}_s .
\end{aligned}
\tag{7.6}
$$

Here, $\kappa_\omega = 2\pi\lambda_\omega|h_\omega|^2/\hbar^2 = (S_{ff}[-\omega] - S_{ff}[\omega])/\hbar^2$ is the system-bath coupling rate at frequency ω.

As our first observation, we note that one can expand the dissipator $\mathcal{D}[\hat{a} + \frac{2\epsilon_2}{\omega_d}\hat{a}^\dagger]$ in Eq. (7.6) to find a *heating* term that remains finite even at *zero temperature*: $\kappa_{\omega_d/2}\left(\frac{2\epsilon_2}{\omega_d}\right)^2\mathcal{D}[\hat{a}^\dagger]$. Its physical origin is a drive photon at frequency ω_d being converted to an oscillator excitation and an environment excitation, both at $\omega_d/2$. The associated effective Unruh-like temperature grows with the squeezing amplitude.

The dominant correction for the situation that interests us, however, is the parity-preserving two-photon heating term $\mathcal{D}[\hat{a}^{\dagger 2}]$. Its physical origin is in the thermal fluctuation at frequency ω_d driving incoherently the parametric process engineered to generate squeezing [53, 66].

With the Lindbladian at order φ_{zps}^1, we can compute the decoherence time T_X of the ground coherent states, otherwise known as Glauber states, of the system $|\alpha = \pm\sqrt{\epsilon_2/K}\rangle \approx |\pm X\rangle$, where X stands for a Bloch sphere axis [53, 66]. This quantity is the smallest non-zero real part of the Lindbladian eigenspectrum [4, 53]. In Fig. 7.1, we plot this quantity as a function of the squeezing amplitude. For our simulation parameters, we take $\kappa_{\omega_d} = 5\ \mu\text{s}^{-1}$ and temperature $T_{\omega_d} = 350$ mK, which are reasonable values for a drive port considering standard couplings and the noise temperature of the electronics controlling the microwave signals in quantum circuit experiments. In addition, we choose $\kappa_{\omega_d/2} = \kappa_{3\omega_d/2} = 0.05\ \mu\text{s}^{-1}$ and temperature $T_{\omega_d/2} = T_{3\omega_d/2} = 50$ mK, which are also based on experimental conditions of interest for this note. The nonlinear coefficients are taken to be $g_3/6\pi = 20$ MHz, $g_4/8\pi = 280$ kHz, and consequently $K/2\pi = 320$ kHz, which are standard values for the SNAIL transmon [52] used in the experiments [53]. The drive frequency is $\omega_d/2\pi = 12$ GHz and the renormalized detuning in Eq. (7.4) is taken to be $\Delta = 0$.

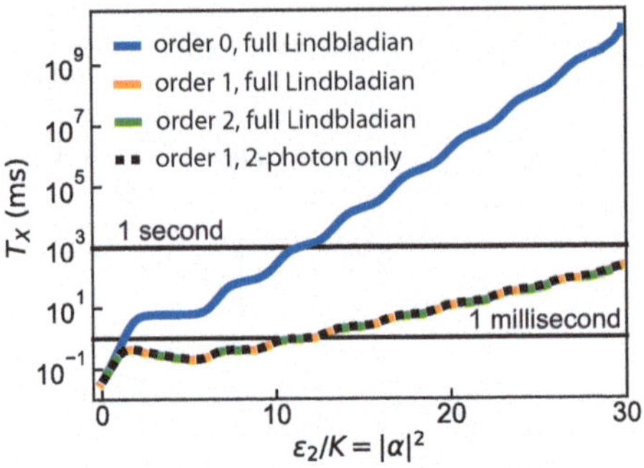

Fig. 7.1 Theoretical simulation of T_X vs ϵ_2/K for different orders of perturbation theory and considering experimentally relevant system and bath parameters. Predictions to order φ_{zps}^0 in the coupling to the environment, corresponding to the ordinary Lindbladian treatment (containing only single photon loss and gain at $\omega = \omega_d/2$), to order φ_{zps}^1, and to order φ_{zps}^2 are shown in blue, orange, and green respectively. Also shown is the effect of keeping only the two-photon terms at order φ_{zps}^1 (see Eq. (7.7), black)

In Fig. 7.1, we show the Lindbladian prediction for the ordinary dissipators (order φ_{zps}^0, blue) and that for dissipators to order φ_{zps}^1 (orange). The two predictions disagree by several orders of magnitude, and thus the former, being incomplete, is unfit to describe state-of-the-art experiments [53]. The prediction to order φ_{zps}^2 (green), which we discuss in detail next, adds negligible corrections and shows the convergence of the method for the chosen parameter values. We note that the ratio of the prefactors of two-photon heating at order φ_{zps}^1 and single photon heating at order φ_{zps}^0 is 17 Hz/12 Hz $\sim$ 1. Yet the two-photon process becomes dominant for $\epsilon_2/K > 2$ because its strength scales as $\langle(\hat{a}^\dagger\hat{a})^2\rangle \sim |\alpha|^4$ while that of the single photon process scales as $\langle\hat{a}^\dagger\hat{a}\rangle \sim |\alpha|^2$. We also plot the Lindbladian prediction (black), computed from

$$\begin{aligned}\partial_t\hat{\rho}_s =& \frac{1}{i\hbar}[\hat{\mathcal{H}}_s, \hat{\rho}_s] + \kappa_{\omega_d/2}\bar{n}_{\omega_d/2}\mathcal{D}[\hat{a}^\dagger]\hat{\rho}_s \\ &+ \kappa_{\omega_d/2}\left(1+\bar{n}_{\omega_d/2}\right)\mathcal{D}[\hat{a}]\hat{\rho}_s \\ &+ \kappa_{\omega_d}\bar{n}_{\omega_d}\left(\frac{8g_3}{3\omega_d}\right)^2\mathcal{D}\left[\hat{a}^{\dagger 2}\right]\hat{\rho}_s \\ &+ \kappa_{\omega_d}\left(1+\bar{n}_{\omega_d}\right)\left(\frac{8g_3}{3\omega_d}\right)^2\mathcal{D}\left[\hat{a}^2\right]\hat{\rho}_s,\end{aligned} \tag{7.7}$$

which only adds to the linear dissipators the term $\propto \mathcal{D}[\hat{a}^{\dagger 2}]$ (and its conjugate). Its close similarity with the full Lindbladian prediction confirms that two-photon heating constitutes the dominant corrections to the ordinary Lindbladian. Note, though, that this decoherence process has only a marginal effect on the lifetime of large Schrödinger cat states ($\propto$ $|\alpha\rangle \pm |-\alpha\rangle$), since it conserves the parity of the state. Despite the failure of the ordinary Lindbladian to predict the lifetime of the coherent states, the lifetime of the Schrödinger cat states measured in [53] is still accounted for by the ordinary linear dissipation because of its inherent fragility to single photon loss events.

Effective Lindbladian at Order φ^2_{zps}: Second Order Beyond the RWA

Similarly to the computation done at order φ^1_{zps}, we also compute $\hat{S}(t)$ generating the unitary transformation Eq. (7.3) to order φ^2_{zps}, as well as the effective Lindbladian to this order. The full expression is given in Eq. (7.16) in the appendix. The correction to this order may become relevant depending on the choice of parameters in the model, as we now discuss.

The second order Lindbladian samples the noise spectrum at $5\omega_d/2$, $2\omega_d$ and near zero frequency in addition to those sampled at the lower orders. For the noise spectrum at these frequencies, we chose $\kappa_{5\omega_d/2} = \kappa_{2\omega_d/2} = 50\ \text{ms}^{-1}$ and $T_{5\omega_d/2} = T_{2\omega_d} = 50$ mK. For zero frequency, we take $\kappa_0 = 0$. These parameter assignations were used also for the calculation to order φ^2_{zps} in Fig. 7.1. We remark that the assignation for κ_0 is an important assumption, justified for the decoherence model proposed here. For a thermal bath of linear oscillators, the number of photons diverges near DC as $\bar{n}_{\text{th}} \sim k_B T/\hbar\omega$ while the density of modes (and thus κ_ω) goes to zero as a polynomial in ω ($\propto \omega^2$ for a resistance coupled to the circuit by a capacitance). Thus, the noise spectral density at near-DC frequency goes to zero as $\omega \to 0$ in this model. However, for other noise models better suited to describe the low-frequency band including, for example quasi-particle loss and inductive loss [105, 123, 145], the noise near DC could become dominant and Eq. (7.2c) should also be extended to capture the corresponding coupling terms.

In Fig. 7.2, we show the effect of increasing the Kerr nonlinearity in the lifetime prediction at order φ^2_{zps} while keeping the rank-three nonlinearity constant to $g_3/6\pi = 20$ MHz as in [53, 66]. The Kerr coefficient is varied by varying g_4 [52]. The dominant dissipator appearing at this order is $\mathcal{D}[\hat{a}^\dagger\hat{a}]$, which on the coherent states acts as a single photon gain enhanced by a factor $|\alpha|^2$. The magnitude of this dissipator scales as $|K|^4|\Pi|^2\langle(\hat{a}^\dagger\hat{a})^2\rangle \propto |K|^4|\alpha|^8$ (see Eq. (7.16)) and its prefactor ranges between 10^{-6} and 2 times that of the dissipator $\mathcal{D}[\hat{a}^{\dagger 2}]$ at order φ^1_{zps} when $K/2\pi$ is varied from 0.25 to 8 MHz for a coherent state with $|\alpha|^2 = 20$. Consequently, this term becomes dominant for $K/2\pi > 2$ MHz and for sufficiently large coherent state amplitudes. This is in qualitative agreement with the fact that the device in [66], characterized by $K/2\pi = 6.7$ MHz, has a T_X lifetime

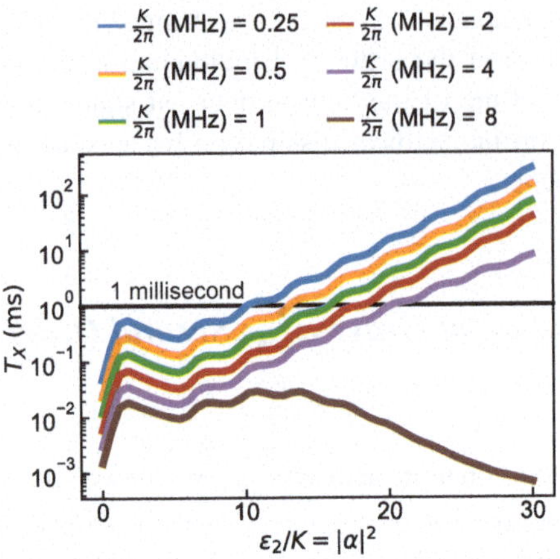

Fig. 7.2 Theoretical predictions for the coherent state lifetime T_X of the Kerr-cat qubit to order φ_{zps}^2 as a function of its mean photon number for several values of the Kerr constant K. The third rank nonlinearity is kept constant for all curves at $g_3/6\pi = 20$ MHz. The bath parameters are identical to Fig. 7.1. The brown curve corresponds to the prediction for parameters within 10% of those in [66], whereas the blue curve corresponds to the prediction for parameters within 10% of those in [53]

considerably lower than the one achieved in [53] where the device was operated at $K/2\pi = 320$ kHz.[1]

The main point of the exploration presented in this chapter is to showcase that an in-depth theoretical understanding of the dissipative processes at various orders is necessary for the experimental activity on parametric processes, like amplification, driven qubits, and quantum gates.

Comparison Between Theoretical Predictions and Experimental Results

We have found that an ordinary Lindbladian treatment is incomplete by several orders of magnitude when the beyond-RWA terms are examined for the Kerr-cat qubit. Consequently, to account for experimental observations, higher orders in the Lindbladian need to be considered. With the analytical expression presented here, we are able to account for the order of magnitude of the observations presented in [53], which are reproduced as maroon dots in Fig. 7.3. Note that for $|\alpha|^2 < 2$,

[1] Note that in order to achieve a given large Kerr ($\gg \kappa_{\omega_d}/2$), and thus fast gates in the Kerr-cat qubit, one should reduce as much as possible the decoherence induced by g_3 and g_4. One sees from the analytical expression in Eq. (7.16) that the prefactors of some dissipators can be minimized or even cancelled, at constant Kerr, by the proper choice of the oscillator's nonlinearities.

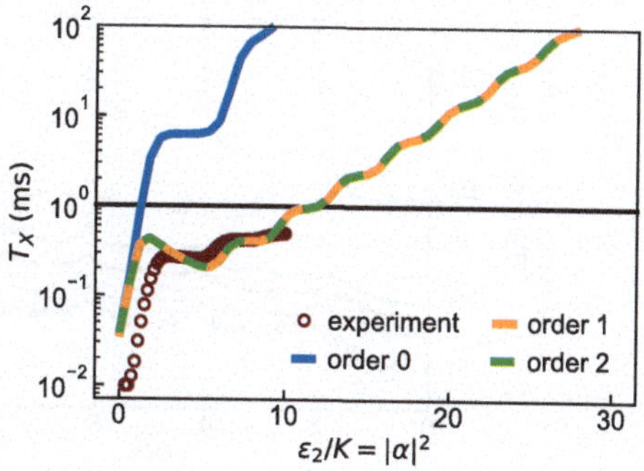

Fig. 7.3 Comparing the staircase experimental data versus predictions of the beyond RWA model. The experimental points correspond to the data in [53]. The bath parameters in the model are identical to Fig. 7.1

where there is a discrepancy between the experimental results and the predictions presented here, the data has been explained in [53], by the inclusion of non-Markovian low-frequency noise which is not included here. The results presented in this chapter emphasise the need for further experiments that will in turn lead to detailed modeling of possible noise sources affecting particularly driven qubits.

Modeling T_X for $\Delta \neq 0$

So far, we only analyzed T_X only as a function of ϵ_2/K for $\Delta = 0$. In this section, we model the transverse relaxation lifetime measurements T_X of the Δ-Kerr-cat qubit for $\Delta \neq 0$. For the sake of simplicity, we only perform simulations over the RWA Lindblad master equation

$$\partial_t \hat{\rho} = \frac{1}{i\hbar}[\hat{H}, \hat{\rho}] + \kappa(1 + \bar{n}_{\text{th}})\mathcal{D}[\hat{a}]\hat{\rho} + \kappa\bar{n}_{\text{th}}\mathcal{D}[\hat{a}^\dagger]\hat{\rho}, \tag{7.8}$$

where $\hat{\rho}$ describes the state of the system, $\bar{n}_{\text{th}} = 1/(\exp(\hbar\omega_a/k_B T) - 1)$ corresponds to the temperature of the environment and κ corresponds to the coupling between system and environment. The Hamiltonian $\hat{H}$ is given by Eq. (3.24) and the dissipator $\mathcal{D}$ of the operator $\hat{O}$ is given by $\mathcal{D}[\hat{O}]\bullet := \hat{O}\bullet\hat{O}^\dagger - (\hat{O}^\dagger\hat{O}\bullet + \bullet\hat{O}^\dagger\hat{O})/2$. In Eq. (7.8), these operators correspond to single photon loss $\mathcal{D}[\hat{a}]$ and gain $\mathcal{D}[\hat{a}^\dagger]$ [16, 21, 22]. In Fig. 7.4, we compare the data presented in Figure 3 of the main text with the lifetime extracted from Eq. (7.8) for different values of n_{th}. The value of κ has been set to $\kappa = 1/T_1 = 1/20\ \mu\text{s}^{-1}$. The current model seems insufficient to accurately predict the observations and more research is needed to understand the decoherence of nonlinear driven systems (see, for example, [152]). Figure 7.4 emphasizes the need for further measurements and a detailed modeling of possible noise sources affecting particularly driven qubits. We also present, in Fig. 7.5, the expected T_X as a function of ϵ_2/K for different values of Δ. This plot indicates that a Δ-Kerr-cat, in general, gives larger T_X lifetimes than a Kerr-cat ($\Delta = 0$).

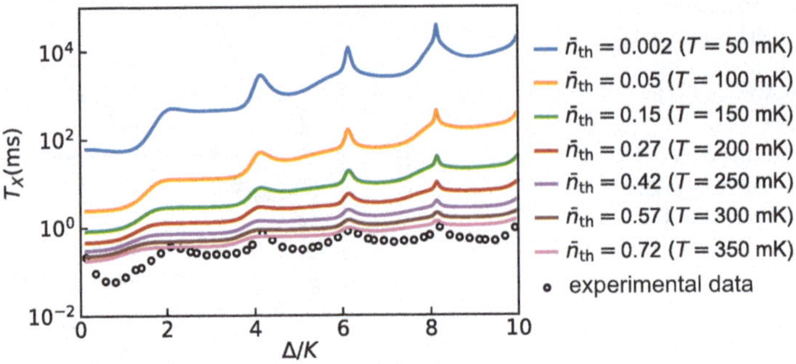

Fig. 7.4 Ordinary Lindblad simulations of T_X as a function of Δ for different thermal populations, corresponding to Eq. (7.8). Black dots correspond to experimental data presented in Fig. 6.5. The value of κ has been taken to be $\kappa = 1/T_1 = 1/20\ \mu\text{s}^{-1}$. The solid curves take the experimentally observed ac Stark shift into account. An ordinary Lindbladian at non-zero temperature is insufficient to predict the experimental data. Beyond-RWA effects may be important to consider [152]

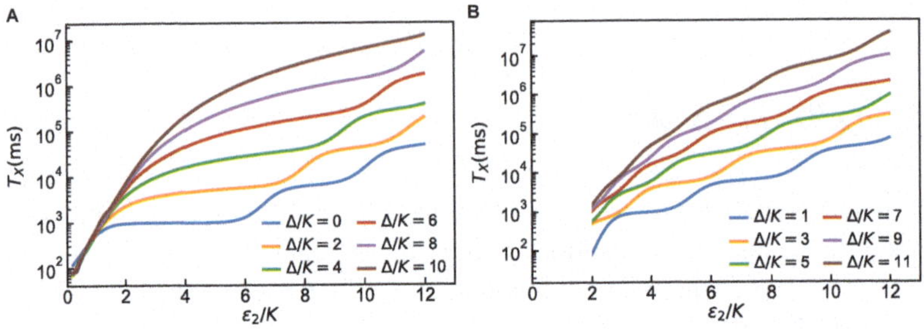

Fig. 7.5 Ordinary Lindblad simulations of T_X as a function of ϵ_2/K for different values of Δ/K, corresponding to Eq. (7.8). For both (**a**) and (**b**), the value of $\kappa/K = 1/50$ and $\bar{n}_{\text{th}} = 0.05$. In (**b**) for $\epsilon_2/K < 2$ the lifetime is limited by ground state tunneling and is this not well captured by our simplified method

Mitigating Lifetime Reduction by Adding Two-Photon Cooling

We identify that the dominant decoherence mechanisms are two-photon heating $\mathcal{D}[\hat{a}^{\dagger 2}]$ and dephasing $\mathcal{D}[\hat{a}^\dagger\hat{a}] \propto |\alpha|^2\mathcal{D}[\hat{a}^\dagger]$. To counteract this, we include, in our computation, a small amount of engineered two-photon dissipation [42, 96, 97, 113, 127] ($\kappa_{2\text{ph}} = 0.003\ \mu\text{s}^{-1}$). We show the outcome of the calculation in Fig. 7.6 for system and bath parameters as in Figs. 7.1 and 7.3. Experimentally, this should be easily achievable since much larger two-photon cooling rates have been demonstrated [97, 150], albeit in absence of a Kerr nonlinearity. Note, however, that a correct understanding is likely to require a higher-order analysis of engineered dissipation [56, 129] like the one presented here. It is likely that the combination of Hamiltonian stabilization and reservoir engineering will provide the agility and fast universal gates for cat-qubits and high coherent state lifetimes [127].

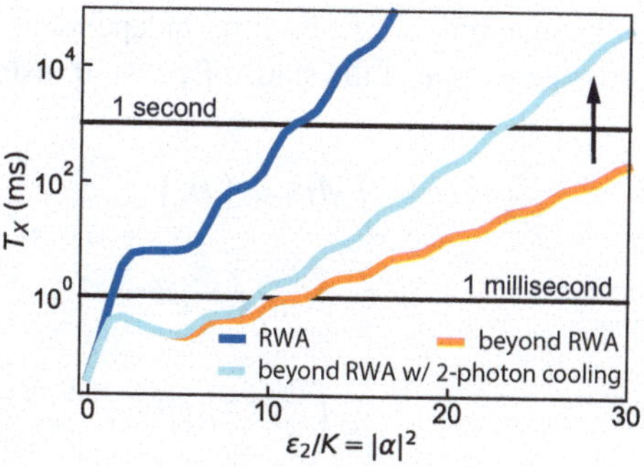

Fig. 7.6 Effect of additional two-photon cooling on T_X. Adding an artificial two-photon dissipation term with a relatively small prefactor to the Kerr-cat system largely compensates the effect of the higher-order dissipators computed here. System and bath parameters have been chosen to be identical to those in Fig. 7.1

Deriving the Static Effective Hamiltonian of the System-Bath Hamiltonian

Here we follow [153] to compute $\hat{S}$ that generates the sought-after canonical transformation. First we expand Eq. (7.3) as

$$\hat{\mathcal{H}}_{\text{eff}} \equiv e^{\hat{S}/i\hbar}\hat{H}_{\text{tot}}(t)e^{-\hat{S}/i\hbar} - i\hbar e^{\hat{S}/i\hbar}\partial_t e^{-\hat{S}/i\hbar} \tag{7.9a}$$

$$\begin{aligned} &= \hat{H}_{\text{s}} + \frac{1}{i\hbar}[\hat{S}, \hat{H}_{\text{s}}] + \frac{1}{2!(i\hbar)^2}[\hat{S}, [\hat{S}, \hat{H}_{\text{s}}]] + \cdots \\ &\quad + \hat{H}_{\text{sb}} + \frac{1}{i\hbar}[\hat{S}, \hat{H}_{\text{sb}}] + \frac{1}{2!(i\hbar)^2}[\hat{S}, [\hat{S}, \hat{H}_{\text{sb}}]] + \cdots \\ &\quad + \partial_t \hat{S} + \frac{1}{2!i\hbar}[\hat{S}, \partial_t \hat{S}] + \cdots \end{aligned} \tag{7.9b}$$

$$= \hat{\mathcal{H}}_{\text{s}} + \hat{\mathcal{H}}_{\text{sb}}, \tag{7.9c}$$

where in Eq. (7.9b) we have plugged in $\hat{H}_{\text{tot}} = \hat{H}_s + \hat{H}_{\text{sb}}$ as defined in Eq. (7.2) and employed the Baker-Campbell-Hausdorff formula; in Eq. (7.9c) $\hat{\mathcal{H}}_{\text{sb}}$ corresponds to the second line of Eq. (7.9b) and $\hat{\mathcal{H}}_{\text{s}}$ consists of the rest of Eq. (7.9b), which contains no bath modes.

Our goal is to perturbatively find $\hat{S}$ so that in the corresponding frame $\hat{\mathcal{H}}_{\text{s}}$ is time-independent to some desired order of φ_{zps}. We therefore write $\hat{H}_{\text{s}}$ and $\hat{S}$ each as a series

$$\hat{H}_{\text{s}} = \sum_{k>0} \hat{H}_{\text{s}}^{(k)}, \quad \hat{S} = \sum_{k>0} \hat{S}^{(k)} \tag{7.10}$$

where $\hat{H}_{\text{s}}^{(k)}$ and $\hat{S}^{(k)}$ are the order φ_{zps}^k components in the corresponding series.

Demanding $\hat{\mathcal{H}}_s$ to be time-independent at order φ_{zps}^1 [153], we obtain the first order generator of the static effective transformation as

$$\frac{\hat{S}^{(1)}}{\hbar} = -\int dt \ \mathrm{osc}\left(\hat{H}_s\right) \tag{7.11}$$

$$\begin{aligned}
&= \frac{2}{5} i \frac{g_3}{\omega_d} a^\dagger \Pi^{*2} e^{i5\omega_d t/2} + \frac{1}{2} i \frac{g_3}{\omega_d} a^{\dagger 2} \Pi^* e^{i4\omega_d t/2} \\
&+ \left(\frac{2}{3} i \frac{g_3}{\omega_d} a \Pi^{*2} + \frac{2}{9} i \frac{g_3}{\omega_d} a^{\dagger 3} \right) e^{i3\omega_d t/2} \\
&+ 2i \frac{g_3}{\omega_d} a^\dagger a \Pi^* e^{i2\omega_d t/2} \\
&+ \left(4i \frac{g_3}{\omega_d} |\Pi|^2 a^\dagger + 2i \frac{g_3}{\omega_d} a^{\dagger 2} a + 2i \frac{g_3}{\omega_d} a^\dagger \right) e^{i\omega_d t/2} \\
&+ \mathrm{h.c.}
\end{aligned} \tag{7.12}$$

where $\mathrm{osc}(f) = f - \int_0^T dt \ f$ extracts the oscillating part of f with T being its periodicity.

At this order, the transformed system-bath coupling is

$$\hat{\mathcal{H}}_{sb}^{(1)} = \frac{[\hat{S}^{(1)}, \hat{H}_{sb}]}{i\hbar}, \tag{7.13}$$

where $\hat{H}_{sb}$ is taken to be of order φ_{zps}^0. Carrying out the calculation explicitly, one then obtains $\hat{\mathcal{H}}_{sb}^{(1)}$ in Eq. (7.5).

At order φ_{zps}^2, the generator of the canonical transformation is accordingly given by

$$\frac{\hat{S}^{(2)}}{\hbar} = -\int_0^t dt \ \mathrm{osc}\left(\hat{H}_s^{(2)} + \frac{[\hat{S}^{(1)}, \hat{H}_s^{(1)}]}{i\hbar} + \frac{[\hat{S}^{(1)}, \partial_t \hat{S}^{(1)}]}{2! i\hbar} \right), \tag{7.14}$$

the system-bath coupling is

$$\begin{aligned}
\hat{H}_{sb}^{(2)} &= \frac{1}{i\hbar} \left[S^{(2)}, \hat{H}_{sb}(t) \right] \\
&+ \frac{1}{2!} \left(\frac{1}{i\hbar} \right)^2 \left[S^{(1)}, \left[S^{(1)}, \hat{H}_{sb}(t) \right] \right],
\end{aligned} \tag{7.15}$$

and the full Lindbladian master equation up to this order is

$$
\begin{aligned}
\partial_t \hat{\rho}_{\mathrm{s}} = & \frac{1}{i\hbar}\left[\hat{\mathcal{H}}_{\mathrm{s}}^{(2)}, \hat{\rho}_{\mathrm{s}}\right] + \kappa_0 (1+\bar{n}_0)\, \mathcal{D}\left[32\frac{g_3^2}{\omega_d^2}\hat{a}^2\Pi^*\right] + \kappa_0 \bar{n}_0\, \mathcal{D}\left[32\frac{g_3^2}{\omega_d^2}\hat{a}^{\dagger 2}\Pi\right] \\
& + \kappa_{\omega_d/2}\left(1+\bar{n}_{\omega_d/2}\right)\left(\mathcal{D}\left[\hat{a} + \frac{2g_3}{\omega_d}a^\dagger\Pi - \left(\frac{35}{2}\frac{g_3^2}{\omega_d^2} - 6\frac{g_4}{\omega_d}\right)\hat{a}|\Pi|^2\right.\right. \\
& \left.\left. - \left(\frac{152}{9}\frac{g_3^2}{\omega_d^2} - 3\frac{g_4}{\omega_d}\right)\hat{a}^\dagger\hat{a}^2 - \left(\frac{152}{9}\frac{g_3^2}{\omega_d^2} - 3\frac{g_4}{\omega_d}\right)\hat{a}\right]\hat{\rho}_{\mathrm{s}}\right) \\
& + \kappa_{\omega_d/2}\bar{n}_{\omega_d/2}\left(\mathcal{D}\left[\hat{a}^\dagger + \frac{2g_3}{\omega_d}\hat{a}\Pi^* - \left(\frac{35}{2}\frac{g_3^2}{\omega_d^2} - 6\frac{g_4}{\omega_d}\right)\hat{a}^\dagger|\Pi|^2\right.\right. \\
& \left.\left. - \left(\frac{152}{9}\frac{g_3^2}{\omega_d^2} - 3\frac{g_4}{\omega_d}\right)\hat{a}^{\dagger 2}\hat{a} - \left(\frac{152}{9}\frac{g_3^2}{\omega_d^2} - 3\frac{g_4}{\omega_d}\right)\hat{a}^\dagger\right]\hat{\rho}_{\mathrm{s}}\right) \\
& + \kappa_{\omega_d}\left(1+\bar{n}_{\omega_d}\right)\mathcal{D}\left[\frac{8g_3}{3\omega_d}\hat{a}^2 - \left(\frac{592}{9}\frac{g_3^2}{\omega_d^2} - 16\frac{g_4}{\omega_d}\right)\hat{a}^\dagger\hat{a}\Pi\right]\hat{\rho}_{\mathrm{s}} \\
& + \kappa_{\omega_d}\bar{n}_{\omega_d}\mathcal{D}\left[\frac{8g_3}{3\omega_d}\hat{a}^{\dagger 2} - \left(\frac{592}{9}\frac{g_3^2}{\omega_d^2} - 16\frac{g_4}{\omega_d}\right)\hat{a}^\dagger\hat{a}\Pi^*\right]\hat{\rho}_{\mathrm{s}} \\
& + \kappa_{3\omega_d/2}\left(1+\bar{n}_{3\omega_d/2}\right)\mathcal{D}\Big[\frac{3g_3}{\omega_d}\hat{a}\Pi - \left(\frac{51}{5}\frac{g_3^2}{\omega_d^2} - \frac{9}{2}\frac{g_4}{\omega_d}\right)\hat{a}^\dagger\Pi^2 \\
& \qquad + \left(4\frac{g_3^2}{\omega_d^2} + \frac{3}{2}\frac{g_4}{\omega_d}\right)\hat{a}^3\Big]\hat{\rho}_{\mathrm{s}} \\
& + \kappa_{3\omega_d/2}\bar{n}_{3\omega_d/2}\mathcal{D}\Big[\frac{3g_3}{\omega_d}\hat{a}^\dagger\Pi^* - \left(\frac{51}{5}\frac{g_3^2}{\omega_d^2} - \frac{9}{2}\frac{g_4}{\omega_d}\right)\hat{a}\Pi^{*2} \\
& \qquad + \left(4\frac{g_3^2}{\omega_d^2} + \frac{3}{2}\frac{g_4}{\omega_d}\right)\hat{a}^{\dagger 3}\Big]\hat{\rho}_{\mathrm{s}} \\
& + \kappa_{2\omega_d}\left(1+\bar{n}_{2\omega_d}\right)\mathcal{D}\left[\left(\frac{224}{45}\frac{g_3^2}{\omega_d^2} + \frac{16}{5}\frac{g_4}{\omega_d}\right)\hat{a}^2\right]\hat{\rho}_{\mathrm{s}} \\
& + \kappa_{2\omega_d}\bar{n}_{2\omega_d}\mathcal{D}\left[\left(\frac{224}{45}\frac{g_3^2}{\omega_d^2} + \frac{16}{5}\frac{g_4}{\omega_d}\right)\hat{a}^{\dagger 2}\right]\hat{\rho}_{\mathrm{s}} \\
& + \kappa_{5\omega_d/2}\left(1+\bar{n}_{5\omega_d/2}\right)\mathcal{D}\left[\left(\frac{19}{9}\frac{g_3^2}{\omega_d^2} + \frac{5}{2}\frac{g_4}{\omega_d}\right)\hat{a}^2\right]\hat{\rho}_{\mathrm{s}} \\
& + \kappa_{5\omega_d/2}\bar{n}_{5\omega_d/2}\mathcal{D}\left[\left(\frac{19}{9}\frac{g_3^2}{\omega_d^2} + \frac{5}{2}\frac{g_4}{\omega_d}\right)\hat{a}^{\dagger 2}\right]\hat{\rho}_{\mathrm{s}}.
\end{aligned}
\tag{7.16}
$$

One can also obtain the photon-number-dependence and Kerr-dependence of relevant terms above using the relationship $g_3\Pi = K|\alpha|^2$ and $K = -3g_4/2 + 20g_3^2/3\omega_d$. With this, one sees that the prefactor of $\mathcal{D}[\hat{a}^\dagger\hat{a}]$ at frequency ω_d is $\propto |K|^4$. Such strong dependence on K explains the drastic drop in T_X for $K/2\pi > 2$ MHz in Fig. 7.2, while for $K/2\pi < 2$ MHz, the effect of $\mathcal{D}[\hat{a}^\dagger\hat{a}]$, which is of order φ_{zps}^2, is much weaker than the effect of dissipators at lower orders and thus the change of the former is masked by that of the latter when K varies in this regime. Note that by engineering the Hamiltonian nonlinearities g_3 and g_4, one may be able to mitigate the effect of these dissipators even for a system with large K.

Refinement of the Model and Further Topics

When deriving the effective Lindbladian, we have made a few important assumptions.

First, we note that we use the usual Born approximation which amounts to assuming $h_j \ll \omega_o\varphi_{\mathrm{zps}}^2$. This is, the Born approximation induces an error $\mathcal{O}(h_j)$ which needs to remain much smaller than the perturbative corrections computed, which are of order φ_{zps}^2 in this work. Under the same assumption, we demand the transformed system-bath Hamiltonian $\hat{\mathcal{H}}_{\mathrm{eff}} = \hat{\mathcal{H}}_{\mathrm{s}} + \hat{\mathcal{H}}_{\mathrm{sb}}$ to be static to order φ_{zps}^2. But since $\hat{\mathcal{H}}_{\mathrm{sb}} = \mathcal{O}(h_j)$, this amounts to demand that only $\hat{\mathcal{H}}_{\mathrm{s}}$ be static, which provides an important but nonessential simplification.

We also remark that, in the standard Born-Markov approximation [21], one treats the system-bath coupling term in the interaction picture, i.e. $\exp(-\hat{\mathcal{H}}_{\mathrm{s}}/i\hbar)\hat{\mathcal{H}}_{\mathrm{sb}}\exp(\hat{\mathcal{H}}_{\mathrm{s}}/i\hbar)$, instead of $\hat{\mathcal{H}}_{\mathrm{sb}}$ as we did in this work. The omission of this frame transformation is valid under the assumption that the bath is white in the neighbourhood of any given frequency ω_j with a width of a few K's wide covering the relevant portion of the spectrum of $\hat{\mathcal{H}}_{\mathrm{s}}$. This assumption holds generally for $\omega_j \gg K$ [21, 22], but should be dealt with delicately for the near-DC noise, which may be treated numerically. Specifically, one can numerically compute the DC system-bath coupling in the interaction picture defined by $\hat{\mathcal{H}}_s$ in Eq. (7.4). This will transform the DC system-bath coupling to a sum of near-DC terms. One can subsequently trace out the bath under the Born-Markov approximation and obtain the effective Lindbladian.

In conclusion, we went beyond the RWA and computed the static effective dissipators for the Kerr-cat system ($\Delta = 0$) discussing possible new effects that may explain experimental data in [53]. Our systematic method, based on [153], can be extended to arbitrary order and can be applied to other controllable driven systems with a residual coupling to a bath.

Chapter 8
Conclusions and Future Directions

Abstract In this chapter, we pose concrete questions that are beyond the scope, but which stem as natural extensions of this thesis work. Both theoretical and experimental questions are posed. Questions are posed in decreasing order of concreteness.

In this thesis work, we presented the experimental realization of a double-well in the deep quantum regime: a squeeze-driven Kerr oscillator (SKO). The SKO corresponds to an effective Hamiltonian model system that is created by submitting a SNAIL superconducting circuit to a rapidly oscillating drive. We experimentally demonstrated multiple, simultaneous degeneracies in the spectrum as a function of the squeeze-drive frequency. Not only can we turn these degeneracies on-and-off on demand, but their number is tunable as a function of the frequency of the squeeze-drive: an omnipresent yet underappreciated knob in driven systems. Importantly, these degeneracies are robust since they are completely independent of the drive amplitude. Moreover, we find that the drive frequency knob not only controls a discrete exact symmetry, that manifests as exact degeneracies, but also controls a continuous approximate symmetry, that manifests as quasi-degeneracies in the ground and excited states of our oscillator. We measured the quasidegeneracies not only as a function of the drive frequency, but also as a function of the drive amplitude in the excited state manifold of our oscillator. These effects combined together and led to our measurement of drastically enhanced well-flip lifetimes culminating in the experimental realization of a super-protected cat qubit in the ground state manifold of our oscillator.

Beyond validating for the first time several proposals from the last three decades [33, 84, 102, 103, 126, 133, 162, 176], our work has important technological applications in quantum annealing [60, 125] and quantum computation and error-correction [60, 63, 137]. A recent flurry of preprints based on our unique experimental setup, yet overlooking the new symmetry we have discovered, have appeared in relation to quantum state tomography [148], excited state phase transitions [27], quantum simulation of spins [30], and quantum simulation of lattice gauge theories [36]. Our system also provides the simplest testbed for parametric processes. Finally, our work

J. Venkatraman, *Controlling the Effective Hamiltonian of a Driven Quantum Superconducting Circuit*, Springer Theses,
https://doi.org/10.1007/978-3-031-83270-3_8

connects the field of Hamiltonian engineering with quantum circuits to the field of "Floquet engineering" [58] with trapped ions and atoms.

Beyond these broad applications, we now pose some concrete questions that are beyond the scope, but which stem as natural extensions of this thesis work. Both theoretical and experimental questions are posed. Questions are posed in decreasing order of concreteness.

8.1 Hidden Symmetries

Question 1
In Eq. (3.1), when Δ equals an even multiple of the Kerr coefficient K, $\Delta/K = 2m$, $m + 1$ exact degeneracies occur in the spectrum, where $m \in \mathbb{Z}_{\geq 0}$, the set of non-negative integers. Importantly, these degeneracies are robust in location and always occur at the same location in Δ and are completely independent of the amplitude ϵ_2. What are the symmetries associated with these exact spectral degeneracies, if any?

We rule out parity symmetry in Fig. 3.6 by plotting the spectrum of another parity-symmetric Hamiltonian $\hat{H} = \Delta\hat{a}^\dagger\hat{a} - K\hat{a}^{\dagger 2}\hat{a}^2 + \epsilon_4(\hat{a}^{\dagger 4} + \hat{a}^4)$, and showing that the location of the expected degeneracies for this system depends on the value of ϵ_4.

Hidden symmetries typically manifest as unexpected, exact spectral degeneracies. Establishing the connection between symmetries and degeneracies in a system is of fundamental importance for the understanding and control of its structural and dynamical properties. This relationship lies at the core of spectacular phenomena like the Jahn-Teller effect [147], the quantum Hall effect [18], and Majorana edge modes [88]. Remarkably, to error correct a quantum computation, the information must be protected by a symmetry that makes the environment blind to it [2, 119]. In this work, we present the discovery of a Kerr parametric oscillator with multiple tunable spectral degeneracies. At the time of the writing of this dissertation, the exact nature of the symmetries, if any, remains an open problem.

8.2 Making a 3-Legged Cat

The theoretical proposal in [126] relied on the fact that the ground states of Eq. (3.21) are degenerate coherent states $|\pm\alpha\rangle$ with $\alpha = \sqrt{\epsilon_2/K}$. Thus, the superpositions of the coherent states, corresponding to the two-legged Schrödinger cat states are also the eigenstates of Eq. (3.21). This is the fundamental idea behind the Kerr-cat qubit. This qubit was experimentally realized in [66]. An immediate extension of the above idea is that the three-legged cat Schrödinger cat states are the ground states of

$$\hat{H} = -K_6\hat{a}^{\dagger 3}\hat{a}^3 + \epsilon_3(\hat{a}^{\dagger 3} + \hat{a}^3), \tag{8.1}$$

which can be seen by the following Hamiltonian factorization

$$\hat{H} = -K_6(\hat{a}^{\dagger 3} - \epsilon_3/K_6)(\hat{a}^3 - \epsilon_3/K_6), \tag{8.2}$$

where in Eq. (8.2) operator-independent terms have been neglected.

From Eq. (8.2), it follows that the two coherent states$|\pm\alpha\rangle$ with $\alpha = \sqrt{\epsilon_3/K_6}$, which are the eigenstates of the annihilation operator $\hat{a}$, are also degenerate eigenstates of Eq. (8.1). The challenge in realizing Eq. (8.1) is twofold. First, the third-order Kerr coefficient K_6 needs to be non-negligible, and of the same order as ϵ_3. Second, to realize coherent states, Eq. (8.2) suggests that lower-order detuning and Kerr-like terms must be perfectly cancelled. These conditions are impossible to achieve in a SNAIL. This is clear by examining Eq. (2.2) and Table 2.1, where $K_6/\omega_o = \mathcal{O}(\varphi_{\text{zps}}^4)$ whereas $K/\omega_o = \mathcal{O}(\varphi_{\text{zps}}^2)$. We propose two solutions to this problem, which leads to two independent future directions.

Question 2

Can we engineer a Josephson circuit with greater control over higher rank nonlinearities $g_k, k \geq 4$, so that higher-order nonlinearities such as K_6 are tunable independent of the lower-order ones? In the experimental work presented in this thesis work, the starting circuit is the SNAIL transmon (or an array of SNAIL transmons) whose only independently controllable Hamiltonian parameters are ω_a, g_3, and g_4. Higher rank nonlinearities are not tunable independently and moreover become progressively weaker. Indeed, adding drives adds an added layer of insitu tunability. By increasing the complexity at the level of circuit design, and then further adding drives, one would add more knobs to tune these nonlinearities, thus potentially making the parameter space of engineerable effective Hamiltonians richer. This is an ongoing and exciting research endeavour in the lab.

Question 3

Can we realize a 3-legged cat in an effective Hamiltonian different from Eq. (8.1)?

By taking inspiration from the fact that adding $\Delta\hat{a}^\dagger\hat{a}$ to Eq. (3.21) still yields degenerate ground states, we answer provide a partial answer to Sect. 8.2 by examining the ground states of the following effective Hamiltonian

$$\hat{H} = \Delta\hat{a}^\dagger\hat{a} - K\hat{a}^{\dagger 2}\hat{a}^2 + \epsilon_3(\hat{a}^{\dagger 3} + \hat{a}^3). \tag{8.3}$$

We provide two proposals below to realize Eq. (8.3):

1. Rapidly drive a SNAIL transmon with drive frequency ω_d configured to be in the vicinity of $3\omega_o/2$,
2. Rapidly drive an ordinary transmon with drive frequency ω_d configured to be in the vicinity of $3\omega_o$.

We elaborate on Property (1) using Appendix B to compute the effective Hamiltonian.

When the SNAIL transmon described in Chap. 2 is submitted to a drive with frequency ω_d in the vicinity of $3\omega_o/2$, the system undergoes a period-tripling

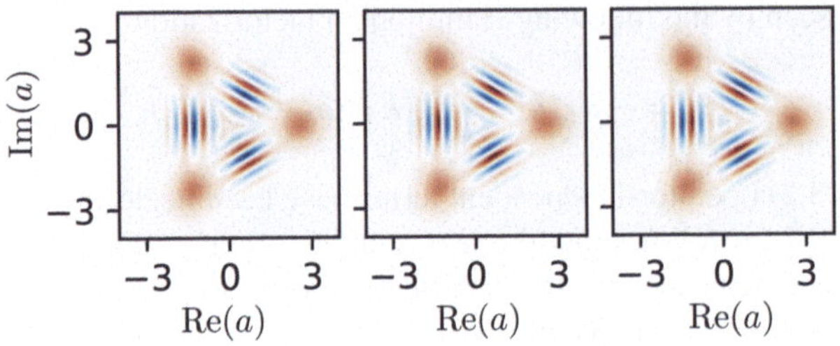

Fig. 8.1 The triply-degenerate ground states of Eq. (8.3)

bifurcation. The effective ground state becomes triply quasi-degenerate and exhibits a three-legged Schrödinger cat manifold (see Fig. 8.1). Under a frame transformation amounting to $\hat{a} \rightarrow \hat{a}e^{-i\frac{2}{3}\omega_d t} + \xi e^{-i\omega_d t} + \mathcal{O}(1/\omega_d)$ where $\xi \approx \frac{9i\Omega_d}{5\omega_d}$, Eq. (2.3) transforms into another time-independent Hamiltonian capturing this effective dynamics

$$\frac{\hat{K}_{\frac{2}{3}\omega_d}}{\hbar} = \Delta\hat{a}^\dagger\hat{a} + \Big(\frac{3g_4}{2} - \frac{5g_3^2}{\omega_d}\Big)\hat{a}^{\dagger 2}\hat{a}^2 + \Big(\frac{195g_3^3}{4\omega_d^2} - \frac{165g_3g_4}{8\omega_d}\Big)\xi^2\hat{a}^{\dagger 3} + \text{h.c.} + \mathcal{O}\left(\frac{1}{\omega_d^3}\right), \tag{8.4}$$

where $\Delta = \omega_o - \frac{2\omega_d}{3} + 3g_4 - \frac{10g_3^2}{\omega_d} + (6g_4 - \frac{180g_3^2}{7\omega_d})|\xi|^2 + \mathcal{O}(\frac{1}{\omega_d^2})$. The threefold symmetry emerges from the "beyond RWA" term $\xi^2\hat{a}^{\dagger 3}$, which, in this condition, is resonant. The three-legged cats emerging from Hamiltonian in the form of Eq. (8.4) maybe used as qutrit. These three-legged cats have received theoretical and experimental attention recently for quantum information processing [99, 175, 176]. In [170], we introduce a diagrammatic approach to draw Feynman-like diagrams and compute the effective Hamiltonian Eq. (8.4).

We leave the simpler exercise of finding the effective Hamiltonian in the case of Property (2) as an exercise for the reader.

8.3 Deviations from the Floquet Quasienergy Spectrum and the Effective Hamiltonian Spectrum

Question 4

When does the spectrum of the effective Hamiltonian diverge from the Floquet quasienergy spectrum [65, 143]? From the calculation presented in Chap. 2, we

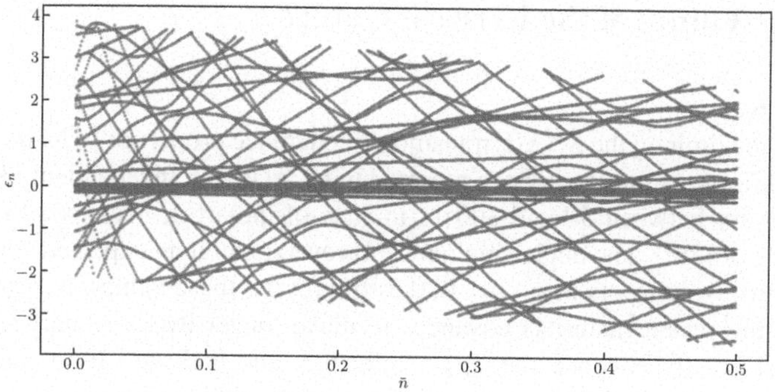

Fig. 8.2 Quasienergies of a driven SNAIL transmon

see that the effective Hamiltonian computed at higher orders deviates from the simple form of a Kerr parametric oscillator Eq. (3.1). Moreover, we expect the driven SNAIL transmon, for progressive drive amplitudes, to exhibit nonlinear resonances [170], quantum diffusion [142], and eventually chaos [34]. However, in the experimental data presented in Fig. 6.1, there is no such signature of any impending chaos. Understanding when the driven SNAIL transmon spectrum shown in Fig. 8.2 starts deviating from the effective Hamiltonian spectrum, may shed light on chaos in driven superconducting circuits, the quantum to classical transition [71, 180, 182], and will form a great testbed for the future of parametric processes in cQED [10, 168].

8.4 Understanding the Well-Flip Lifetime of the Kerr-cat Qubit and Mitigating Unwanted Heating with Engineered Two-Photon Dissipation

Question 5

Find a quantitative model to explain the peaks (see Fig. 6.5) and the staircase (see Fig. 6.2) in the well-flip time of the Kerr-cat qubit. In Chap. 7, we developed a model to explain the several orders of magnitude discrepancy between the observed coherent state lifetime and the predictions of the RWA model. If these beyond RWA effects are at play, filtering the drive lines and ensuring that they are cold will cause a dramatic increase in T_X. This is an experimentally testable prediction. Moreover, by varying the external flux, g_3 can be varied and thus the predictions shown in Fig. 7.2 can be tested too. Finally, Fig. 7.6 shows that engineered two-photon dissipation can counteract the decoherence induced by two-photon heating, an unfortunate circumstance of driving alone. Moreover, adding controlled two-photon cooling provides an extra knob to play with and thus a richer parameter space to uncover new physics.

8.5 The Future of the Kerr-cat Qubit?

Question 6
What sets the limit on the SNAIL transmon's coherence properties? The decay times T_1 of SNAIL transmons have been reported to be relatively independent of flux, but generally vary between 50 and 100 μs in state-of-the-art devices (see Figure 9 in Appendix I of [26]). The dephasing rates, however, exhibit an expected suppression near the flux sweet spots at $\varphi_{\text{ext}} = 0, 0.5$. Improving these values, by for instance incorporating recent materials advances to make longer-lived ordinary transmons [122], will greatly enhance the T_{YZ} in Fig. 6.3, and thus will ultimately dictate the viability of the Kerr-cat qubit as an ancilla for scalable quantum technologies. Another challenge is the demonstration of gates on two Kerr-cat qubits. Proposals on a topological CNOT gate can be found in refs [49, 128]. Beyond testing the viability of the Kerr-cat qubit for scalable quantum technologies, coupling two SKOs has applications in the study of chaos [61] and the quantum to classical transition [73, 131, 179–182].

Appendix A
Notation

In this work, we note $\hat{X}$ and $\hat{P}$ the *position*-like and *momentum*-like coordinates with $[\hat{X}, \hat{P}] = i\hbar$. We build the dimensionless quadratures by introducing the zero point spread of the coordinates as X_{zps} and P_{zps}, respecting $X_{\text{zps}}P_{\text{zps}} = \hbar/2$. We further introduce the complex notation for the dimensionless quadratures as $\hat{a} = (\hat{X}/X_{\text{zps}} + i\hat{P}/P_{\text{zps}})/2$ and its conjugate operator $\hat{a}^\dagger$, where $[\hat{a}, \hat{a}^\dagger] = 1$ and introduce the rescaled phase space quadratures as $\hat{x} = \sqrt{\lambda/2}\hat{X}/X_{\text{zps}} = \sqrt{\lambda/2}(\hat{a} + \hat{a}^\dagger)$ and $\hat{p} = \sqrt{\lambda/2}\hat{P}/P_{\text{zps}} = -i\sqrt{\lambda/2}(\hat{a} - \hat{a}^\dagger)$, where $[\hat{x}, \hat{p}] = i\lambda$. These choices induce the definitions $x_{\text{zps}} = p_{\text{zps}} = \sqrt{\lambda/2}$. Conversely, we have $\hat{a} = (\hat{x} + i\hat{p})/\sqrt{2\lambda}$. At this point, λ is a dimensionless rescaling parameter. We will connect it with the Hamiltonian parameters later, while discussing the classical limit ($\lambda \to 0$) of our system, and thereby giving it physical significance. It is also useful to compare our results with those of [103], who have performed a WKB analysis of a driven oscillator. Thus, unless otherwise specified, λ should be taken equal to unity $\lambda = 1$.

For a mechanical oscillator with mass m and spring-constant k, the small-oscillation frequency is $\omega_o = \sqrt{k/m}$ and the impedance is $Z_o = 1/\sqrt{km}$. With this, we have $X_{\text{zps}} = \sqrt{\hbar Z_o/2}$ and $P_{\text{zps}} = \sqrt{\hbar/2Z_o}$. We further remark that there is a direct correspondence between the mechanical harmonic oscillator and a linear LC circuit oscillator [10, 40, 57] under the following relations. The mechanical position coordinate $\hat{X}$ corresponds to the circuit flux $\hat{\Phi}$, the mechanical momentum $\hat{P}$ corresponds to the circuit charge $\hat{Q}$, where $[\hat{\Phi}, \hat{Q}] = i\hbar$, the mechanical oscillator frequency $\omega_o = \sqrt{k/m}$ corresponds to the circuit oscillator frequency $\omega_o = 1/\sqrt{LC}$ and the mechanical oscillator impedance $Z_o = 1/\sqrt{km}$ corresponds to the circuit oscillator impedance $Z_o = \sqrt{L/C}$ which amounts to the identification of the mechanical mass m with the circuit capacitance C and the spring constant k with the inverse inductance $1/L$. The expressions for the zero point spreads are given by $\Phi_{\text{zps}} = \sqrt{\hbar Z_o/2}$ and $Q_{\text{zps}} = \sqrt{\hbar/2Z_o}$. In circuits, it is customary to introduce [10, 38] the reduced flux and charge coordinates: $\hat{\varphi} = \sqrt{\lambda}2\pi\hat{\Phi}/\Phi_0$

J. Venkatraman, *Controlling the Effective Hamiltonian of a Driven Quantum Superconducting Circuit*, Springer Theses,
https://doi.org/10.1007/978-3-031-83270-3

Table A.1 Correspondence between a mechanical oscillator and a circuit oscillator

Mechanical oscillator	Circuit oscillator
$\hat{X};\ \hat{P}$	$\hat{\Phi};\ \hat{Q}$
$[\hat{X},\hat{P}]=i\hbar$	$[\hat{\Phi},\hat{Q}]=i\hbar$
$\omega_o=\sqrt{k/m}$	$\omega_o=1/\sqrt{LC}$
$Z_o=1/\sqrt{km}$	$Z_o=\sqrt{L/C}$
$X_{\mathrm{zps}}=\sqrt{\hbar Z_o/2}$; $P_{\mathrm{zps}}=\sqrt{\hbar/2Z_o}$ $\Rightarrow X_{\mathrm{zps}}P_{\mathrm{zps}}=\hbar/2$	$\Phi_{\mathrm{zps}}=\sqrt{\hbar Z_o/2}$; $Q_{\mathrm{zps}}=\sqrt{\hbar/2Z_o}$ $\Rightarrow \Phi_{\mathrm{zps}}Q_{\mathrm{zps}}=\hbar/2$
$\hat{a}=\frac{1}{2}\left(\frac{\hat{X}}{X_{\mathrm{zps}}}+i\frac{\hat{P}}{P_{\mathrm{zps}}}\right)$	$\hat{a}=\frac{1}{2}\left(\frac{\hat{\Phi}}{\Phi_{\mathrm{zps}}}+i\frac{\hat{Q}}{Q_{\mathrm{zps}}}\right)$
$\hat{X}=X_{\mathrm{zps}}\left(\hat{a}+\hat{a}^\dagger\right)$	$\hat{\Phi}=\Phi_{\mathrm{zps}}\left(\hat{a}+\hat{a}^\dagger\right)$
$\hat{P}=-i P_{\mathrm{zps}}\left(\hat{a}-\hat{a}^\dagger\right)$	$\hat{Q}=-i Q_{\mathrm{zps}}\left(\hat{a}-\hat{a}^\dagger\right)$
$\left[\hat{a},\hat{a}^\dagger\right]=1$	$\left[\hat{a},\hat{a}^\dagger\right]=1$
$\hat{x}=\sqrt{\frac{\lambda}{2}}\frac{\hat{X}}{X_{\mathrm{zps}}}=x_{\mathrm{zps}}\left(\hat{a}+\hat{a}^\dagger\right)$	$\hat{\varphi}=\sqrt{\lambda}2\pi\frac{\hat{\Phi}}{\Phi_0}=\varphi_{\mathrm{zps}}\left(\hat{a}+\hat{a}^\dagger\right)$
$\hat{p}=\sqrt{\frac{\lambda}{2}}\frac{\hat{P}}{P_{\mathrm{zps}}}=-ip_{\mathrm{zps}}\left(\hat{a}-\hat{a}^\dagger\right)$	$\hat{N}=\sqrt{\lambda}\frac{\hat{Q}}{2e}=-iN_{\mathrm{zps}}\left(\hat{a}-\hat{a}^\dagger\right)$
$x_{\mathrm{zps}}=p_{\mathrm{zps}}=\sqrt{\lambda/2}$ $\Rightarrow x_{\mathrm{zps}}p_{\mathrm{zps}}=\lambda/2$	$\varphi_{\mathrm{zps}}=2\pi\sqrt{\lambda}\frac{\Phi_{\mathrm{zps}}}{\Phi_0}$; $N_{\mathrm{zps}}=\sqrt{\lambda}\frac{Q_{\mathrm{zps}}}{2e}$ $\Rightarrow \varphi_{\mathrm{zps}}N_{\mathrm{zps}}=\lambda/2$
$[\hat{x},\hat{p}]=i\lambda$	$[\hat{\varphi},\hat{N}]=i\lambda$
$\hat{a}=\frac{1}{2}\left(\frac{\hat{x}}{x_{\mathrm{zps}}}+i\frac{\hat{p}}{p_{\mathrm{zps}}}\right)$	$\hat{a}=\frac{1}{2}\left(\frac{\hat{\varphi}}{\varphi_{\mathrm{zps}}}+i\frac{\hat{N}}{N_{\mathrm{zps}}}\right)$
$\hat{a}=(\hat{x}+i\hat{p})/\sqrt{2\lambda}$	$\hat{a}=\left(\sqrt{\frac{\lambda}{2}}\frac{\hat{\varphi}}{\varphi_{\mathrm{zps}}}+i\sqrt{\frac{\lambda}{2}}\frac{\hat{N}}{N_{\mathrm{zps}}}\right)/\sqrt{2\lambda}$

and $\hat{N}=\sqrt{\lambda}\hat{Q}/2e$ so that $[\hat{\varphi},\hat{N}]=i\lambda$, where e is the charge quantum, and $\Phi_0=h/2e$ is the magnetic flux quantum.[1] Their respective zero point spreads $\varphi_{\mathrm{zps}}=\sqrt{\lambda}2\pi\Phi_{\mathrm{zps}}/\Phi_0$ and $N_{\mathrm{zps}}=\sqrt{\lambda}Q_{\mathrm{zps}}/2e$, and are related to the rescaled complex coordinate operators by $\hat{\varphi}=\varphi_{\mathrm{zps}}(\hat{a}^\dagger+\hat{a})$ and $\hat{N}=-iN_{\mathrm{zps}}(\hat{a}-\hat{a}^\dagger)$ and $\varphi_{\mathrm{zps}}N_{\mathrm{zps}}=\lambda/2$. We summarize this notation in Table A.1 on the next page.

[1] Note that in this case the non-dimensionalization of variables is done by fundamental constants and not by linear properties of the oscillator. This comes at the price of a slight notation asymmetry over the reduced operators the electric and mechanical oscillators.

Appendix B
A Recursive Formula to Compute the Static Effective Hamiltonian of a Rapidly Driven Nonlinear System

We start with the equations governing time evolution of the classical or quantum state vector ρ under the action of a time-dependent Hamiltonian $H(t)$ that we write jointly as

$$\partial_t \rho = \{H, \rho\}, \tag{B.1}$$

where the double bracket can be understood as

$$\{\!\{H, \rho\}\!\} \to \begin{cases} \{H, \rho\} & \text{classical (Liouville)}, \\ \frac{1}{i\hbar}[\hat{H}, \hat{\rho}] & \text{quantum (von Neumann)}. \end{cases} \tag{B.2}$$

Here, we have adopted the standard notation $\{\Box, \Box\}$ for the Poisson bracket over phase-space coordinates x and p and $[\Box, \Box]$ for the Hilbert space commutator. The state vector ρ can be taken to be either a phase-space distribution $\rho(x, p)$ or the density operator $\hat{\rho} = \sum_{x',x''} \rho_{x'x''} |x'\rangle\langle x''|$. Its time evolution is governed by the Hamiltonian H which is either the phase-space Hamiltonian $H(x, p, t)$ or the operator $\hat{H}(x, p, t)$. We note that one can also interpret $\{\Box, \Box\}$ as the Moyal bracket [35, 67, 115, 173], in which case Eq. (B.1) describes the dynamics of the phase-space Wigner distribution.

In this formalism agnostic to the nature of the system, we seek a canonical transformation $\rho \to \varrho$ such that the time evolution of ϱ is governed, in the transformed frame, by the sought-after time-independent Kamiltonian. We thus consider the Lie transformation generated by a time-dependent generator S and parametrized by ϵ,

J. Venkatraman, *Controlling the Effective Hamiltonian of a Driven Quantum Superconducting Circuit*, Springer Theses,
https://doi.org/10.1007/978-3-031-83270-3

$$\begin{aligned} \varrho = e^{\epsilon L_S}\rho = \sum_{k=0} \frac{\epsilon^k L_S^k}{k!}\rho; \\ = \rho + \epsilon\{S,\rho\} + \frac{\epsilon^2}{2!}\{S,\{S,\rho\}\} + \cdots , \end{aligned} \tag{B.3}$$

where $L_S\square = \{S, \square\}$ is the Lie derivative [20, 43, 62, 135, 141] generated by S. Here, S is either a real phase-space function $\tilde{S}(q, p, t)$ or an Hermitian operator $\hat{S}(q, p, t)$. Equivalently, the transformed state ϱ is the solution to the differential equation $\partial_\epsilon \varrho = \{S, \varrho\}$, with initial condition $\varrho(\epsilon = 0) = \rho$.

In the transformed representation, the dynamics obeys formally (B.1) as $\partial_t \varrho = \{\mathcal{K}, \varrho\}$, with the Kamiltonian $\mathcal{K}$ given by

$$\mathcal{K} = e^{L_S} H + \int_0^1 d\epsilon\, e^{\epsilon L_S} \dot{S}; \tag{B.4}$$

see Phys Rev Lett 129(10):100601 Supplement Section A for the derivation. Note that in the quantum case, Eq. (B.4) reduces to the familiar expression [164]:

$$\hat{\mathcal{K}} = \hat{U}^\dagger (\hat{H} - i\hbar\partial_t)\hat{U} \tag{B.5}$$

with $\hat{U} = e^{-\hat{S}/i\hbar}$.

We now carry out a perturbative expansion generated by S, while imposing that $\mathcal{K}$ is rendered time-independent. The transformation of the time evolution from $\rho \to \varrho$ is represented schematically in Fig. B.1 and yields

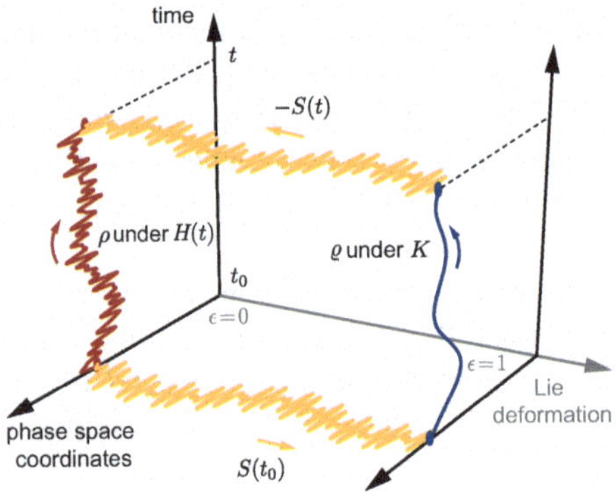

Fig. B.1 Time evolution of the state vector in the transformed ($\epsilon = 1$) and un-transformed ($\epsilon = 0$) frames. The red curve represents the complicated time evolution of ρ under the time-dependent H. The blue curve represents the simpler time evolution of ϱ under the time-independent $\mathcal{K}$. The transformation is exact. Under a sufficiently fast oscillating drive, the fast micromotion captured by S can be neglected and $\mathcal{K}$ can be taken to generate the time evolution of ρ in the un-transformed frame

$$\begin{aligned} \rho &= \mathcal{T} e^{\int_{t_0}^{t} dt' \, L_{H(t')}} \rho_0 \\ &= e^{L_{-S(t)}} e^{L_{\mathcal{K}}(t-t_0)} e^{L_{S(t_0)}} \rho_0, \end{aligned} \tag{B.6}$$

where $\mathcal{T}$ is the time-ordering operator and ρ_0 is the initial state. The time evolution of ρ under H (a Lie transformation generated by H and parametrized by t) can be understood as being decomposed into three successive Lie transformations generated by $S(t_0)$, $\mathcal{K}$, and $-S(t)$. Under this decomposition, the time-ordering operator drops out in the time evolution under $\mathcal{K}$, providing an important simplification.

To carry out the perturbative expansion, we consider the Hamiltonian

$$H(t) = \sum_{n \in \mathbb{N}} \sum_{m \in \mathbb{Z}} H_m^{(n)} e^{im\omega t} \tag{B.7}$$

with period $T = 2\pi/\omega$, and the nth terms to be of order n in the perturbation parameter, here taken to be φ_{zps}. For the perturbative treatment to be valid, the rate of evolution under any one $H_m^{(n)}$ needs to be much smaller than $\omega_d/2$. In the case of an unbounded Hamiltonian, either quantum or classical, the corresponding space will require truncation. We focus on the case of a periodic drive for simplicity, but we note that our treatment can be generalized to include quasiperiodic or non-monochromatic drives; see Phys Rev Lett 129(10):100601 Supplement Section A III in [153] for a concrete example. We take the following ansatz for S and $\mathcal{K}$:

$$S = \sum_{n \in \mathbb{N}} S^{(n)}, \qquad \mathcal{K} = \sum_{n \in \mathbb{N}} \mathcal{K}^{(n)}, \tag{B.8}$$

where we take $S^{(0)} = 0$ and the nth terms to be of order n in the perturbation parameter, here taken to be φ_{zps}. Substituting Eqs. (B.8) into Eq. (B.4) separates the problem into orders of φ_{zps}. At each order, $\mathcal{K}^{(n)}$ can further be expressed as a sum of terms generated by a Lie series as in Eq. (B.3), which we write as $\mathcal{K}^{(n)} = \sum_k \mathcal{K}_{[k]}^{(n)}$. Demanding $\mathcal{K}$ to be time-independent to all orders, we find, after a few lines of algebra, the following coupled recursive formulas:

$$\mathcal{K}_{[k]}^{(n)} = \begin{cases} H^{(n)} & k = 0 \\ \dot{S}^{(n)} + L_{S^{(n)}} H & k = 1 \\ \sum_{m=0}^{n-1} \frac{1}{k} L_{S^{(n-m)}} \mathcal{K}_{[k-1]}^{(m)} & 1 < k \leq n+1 \\ 0 & \text{otherwise,} \end{cases} \tag{B.9a}$$

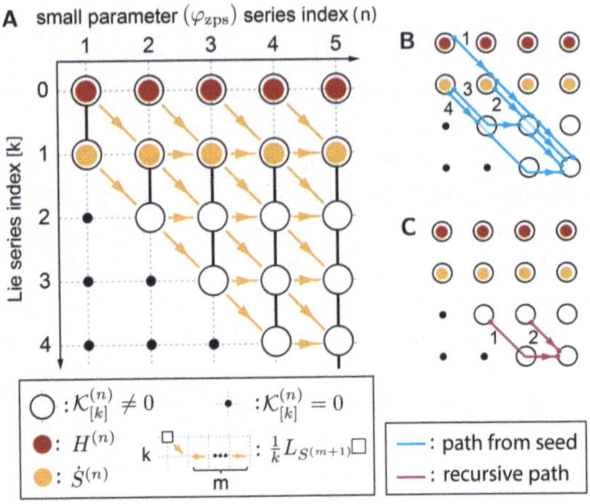

Fig. B.2 (**a**) Colored circles represent the seeds generating the series to all orders. (**b**) As an example, all the paths contributing to the calculation of $\mathcal{K}^{(3)}_{[3]}$ are highlighted. (**c**) Here, only the subpaths contributing to the recursive expression of the aforementioned term are highlighted

$$S^{(n)} = \begin{cases} -\int dt\, \mathbf{osc}(H^{(n)}) & n = 1 \\ -\int dt\, \mathbf{osc}\Big(L_{S^{(n)}} H \\ \quad + \sum\limits_{k>1}^{n+1} \sum\limits_{m=0}^{n-1} \frac{1}{k} L_{S^{(n-m)}} \mathcal{K}^{(m)}_{[k-1]}\Big) & n > 1, \end{cases} \tag{B.9b}$$

where $\mathbf{osc}(f) := f - \overline{f}$, and $\overline{f} = \frac{1}{T}\int_0^T dt\, f$. Note that H is taken to be of order zero in the perturbation parameter, but this hypothesis can be relaxed in a more elaborate treatment; see Phys Rev Lett 129(10):100601 Supplement Section A III in [153] for a concrete example. Note that the difference between Eq. (B.9a) and Fig. B.2 and the corresponding Eq. (8) and Figure 2 in [153] stems from the difference in small parameter. The former is a perturbative expansion in φ_{zps} while the latter is an expansion in inverse powers of the drive frequency (also called a high-frequency or $1/\omega$ expansion).

By construction, taking the time-derivative of Eq. (B.9b), substituting the result into Eq. (B.9a), and summing over k yields a time-independent $\mathcal{K}^{(n)}$. All in all, the computations of $\mathcal{K}$ and S are interleaved so that the computation of $\mathcal{K}^{(n)}$ requires as an input the value of $S^{(m\leq n)}$. Demanding the time-independence of $\mathcal{K}^{(n)}$ fixes $\dot{S}^{(n+1)}$, allowing the recursion to be carried out to the next order. The coupled recursive formula in Eq. (B.9) constructs, as announced, S and $\mathcal{K}$ order-by-order.

The mathematical structure of the recursive formula Eq. (B.9) is shown diagrammatically in Fig. B.2, as we now explain. The figure consists of a grid indexed by the integers n and k. The grid supports a graph. Each node (n, k) of the graph corresponds to a summand $\mathcal{K}^{(n)}_{[k]}$, and the colored ones represent the "seeds" of the calculation. The summand $\mathcal{K}^{(n)}_{[k]}$ is itself a sum of terms, each corresponding to a path connecting the node (n, k) to a seed. Evaluating a path corresponds to taking Lie derivatives over H or $\dot{S}^{(n+1)}$ as dictated by the seed color. The rule is that each Lie derivative is specified by a valid subpath, which must start "downwards" and, when followed by m horizontal edges at row k, contributes with $L_{S^{(m+1)}}/k$. Finally, if the considered node is itself colored, either H or $\dot{S}^{(n+1)}$ must be added to the sum. We note that our grid construction is inspired by [37], where the construction is limited to completely classical and time-independent systems.

Let us discuss, as an example, how $\mathcal{K}^{(3)}_{[3]}$ is evaluated from the figure. As indicated by panel Fig. B.2b, $\mathcal{K}^{(3)}_{[3]}$ contains only four terms corresponding to the concatenations of the valid subpaths (in blue). The sum reads

$$\mathcal{K}^{(3)}_{[3]} = \frac{L_{S^{(1)}}}{1}\frac{L_{S^{(1)}}}{2}\frac{L_{S^{(1)}}}{3}H + \frac{L_{S^{(1)}}}{2}\frac{L_{S^{(1)}}}{3}\dot{S}^{(2)}$$
$$+ \frac{L_{S^{(1)}}}{2}\frac{L_{S^{(2)}}}{3}\dot{S}^{(1)} + \frac{L_{S^{(2)}}}{2}\frac{L_{S^{(1)}}}{3}\dot{S}^{(1)},$$

where the terms are ordered as enumerated in the figure.

Alternatively, one could have expressed $\mathcal{K}^{(3)}_{[3]}$ recursively by directly applying Eq. (B.9a). The computation of $\mathcal{K}^{(3)}_{[3]}$ then involves only the two pink subpaths shown in Fig. B.2c and yields

$$\mathcal{K}^{(3)}_{[3]} = \frac{L_{S^{(2)}}}{3}\mathcal{K}^{(1)}_{[2]} + \frac{L_{S^{(1)}}}{3}\mathcal{K}^{(2)}_{[2]}.$$

At this stage, once all entries of the nth column are computed, the calculation proceeds by demanding the time-independence of $\mathcal{K}^{(n)}$ computed as their row-sum over column n and represented by the vertical bold lines in Fig. B.2a. This is required by Eq. (B.9b). For the column $n = 3$ the algorithm yields

$$S^{(4)} = -\int dt\, \mathbf{osc}\left(L_{S^{(3)}}H + K^{(3)}_{[2]} + K^{(3)}_{[3]} + \mathcal{K}^{(3)}_{[4]}\right),$$

which is a necessary ingredient to compute $\mathcal{K}^{(5)}$ and so, the calculation proceeds.

Appendix C
Computing the Effective Hamiltonian of the Kapitza Pendulum

We take as an illustration of our agnostic formulation the case of a rigid pendulum of length l and mass m whose pivot undergoes a sinusoidal motion of amplitude r along the vertical. This driven pendulum, known as the Kapitza pendulum [86, 94], serves as a model for the dynamical stabilization of mechanical systems, and is described by the classical Hamiltonian

$$H(t) = \frac{p_\varphi^2}{2J} - J\omega_o^2 \cos\varphi - J\frac{r}{l}\omega^2 \cos\varphi \cos\omega t, \tag{C.1}$$

where $J = ml^2$ is the moment of inertia and $\omega_o = \sqrt{\frac{g}{l}}$ is its small oscillation frequency. In Eq. (C.1), φ is the angle between the pendulum and the vertical and p_φ is the angular momentum so that $\{\varphi, p_\varphi\} = 1$. Decomposing the Hamiltonian following the notation in Eq. (6) we have

$$H_0 = \frac{p_\varphi^2}{2J} - J\omega_o^2 \cos\varphi, \qquad H_{\pm 1} = -J\frac{r}{2l}\omega^2 \cos\varphi. \tag{C.2}$$

Usual treatments of the Kapitza pendulum are performed on its equations of motion [94, 130], which are separated into slow and fast coordinates. Averaging over the latter yields an effective potential in the slow coordinates. Here, we recover well-known results following our approach that allows to extend the calculation to higher orders. To demonstrate the coupled construction of S and $\mathcal{K}$ in Eq. (B.9a), we proceed order-by-order. The leading order reads

$$\mathcal{K}^{(0)} = H + \dot{S}^{(1)}. \tag{C.3}$$

By demanding $S^{(1)}$ to be a primitive of $-\,\mathbf{osc}(H)$, i.e. $S^{(1)} = -\int dt\, \mathbf{osc}(H)$, we find $\mathcal{K}^{(0)}$ to be simply the time-average of H:

J. Venkatraman, *Controlling the Effective Hamiltonian of a Driven Quantum Superconducting Circuit*, Springer Theses,
https://doi.org/10.1007/978-3-031-83270-3

$$\begin{aligned}\mathcal{K}^{(0)} &= \overline{H} = \frac{1}{T}\int_0^T H(t)dt \\ &= \frac{p_\varphi^2}{2J} - J\omega_o^2\cos\varphi.\end{aligned} \tag{C.4}$$

Note that, for simplicity, we have set the integration constant of $S^{(1)}$ to be zero and so we shall do for higher orders.

To order one, we have

$$\mathcal{K}^{(1)} = \dot{S}^{(2)} + L_{S^{(1)}}H + \frac{1}{2}L_{S^{(1)}}\dot{S}^{(1)}. \tag{C.5}$$

By demanding $S^{(2)} = -\int dt\, \mathbf{osc}(L_{S^{(1)}}H + \frac{1}{2}L_{S^{(1)}}\dot{S}^{(1)})$, we find

$$\mathcal{K}^{(1)} = -\overline{\left\{\int dt\, \mathbf{osc}(H), \mathbf{osc}(H)\right\}} = 0, \tag{C.6}$$

where $\{\Box_1, \Box_2\}$ in Eq. (C.6) is the Poisson bracket between phase space functions $\Box_1$ and $\Box_2$.

Following this procedure, we compute

$$\mathcal{K}^{(2)} = -J\frac{r^2}{l^2}\frac{\omega^2}{8}\cos 2\varphi. \tag{C.7}$$

For the perturbative treatment to be valid the typical rate of evolution under any one of the terms in Eq. (C.1) must be small with respect to the perturbation frequency ω. The typical rate of evolution under H_0 is naturally ω_o while that under $H_{\pm 1}$ is $\sim \sqrt{\frac{r}{2l}}\,\omega$ and, thus, the perturbative conditions read $\omega_o \ll \omega$ and $r \ll 2l$.

We observe that when the second order correction Eq. (C.7) becomes comparable to the potential energy contribution of the zeroth order, the system develops a secondary stable position at $\varphi = \pi$ [86, 94]. This happens for the well-known condition $\frac{r}{l}\frac{\omega}{\omega_o} > \sqrt{2}$.

Following this approach it is easy to carry the calculation further. Up to order four we find

$$\begin{aligned}\mathcal{K}^{(3)} &= 0 \\ \mathcal{K}^{(4)} &= \frac{r^2}{l^2}\frac{3}{8J}p_\varphi^2(1-\cos 2\varphi) + \frac{r^2}{l^2}\frac{J\omega_o^2}{4}(\cos\varphi - \cos 3\varphi).\end{aligned} \tag{C.8}$$

We see that the fourth order correction introduces a nontrivial modulation to the potential but that no interesting additional stable solution arise due to the perturbative hierarchy of the prefactors.

The quantum calculation proceeds analogously by changing the Lie bracket, i.e., the sub-routine in the algorithm. In this case, the Hamiltonian is given by Dirac's canonical quantization recipe [41] and reads

$$\hat{H}(t) = \frac{\hat{p}_\varphi^2}{2J} - J\omega_o^2 \cos\hat{\varphi} - \frac{r}{l} J\omega^2 \cos\hat{\varphi}\cos\omega t, \tag{C.9}$$

where $\frac{1}{i\hbar}[\hat{\varphi}, \hat{p}_\varphi] = 1$.

To order four the Kamiltonian reads

$$\begin{aligned}
\hat{\mathcal{K}}^{(0)} &= \frac{\hat{p}_\varphi^2}{2J} - J\omega_o^2 \cos\hat{\varphi} \\
\hat{\mathcal{K}}^{(1)} &= 0 \\
\hat{\mathcal{K}}^{(2)} &= -\frac{r^2}{l^2}\frac{J\omega^2}{8}\cos 2\hat{\varphi} \\
\hat{\mathcal{K}}^{(3)} &= 0 \\
\hat{\mathcal{K}}^{(4)} &= \frac{r^2}{l^2}\frac{3}{8J}\hat{p}_\varphi^2(1-\cos 2\hat{\varphi}) + i\hbar\frac{r^2}{l^2}\frac{3}{4J}\hat{p}_\varphi \sin 2\hat{\varphi} \\
&\quad - \hbar^2\frac{r^2}{l^2}\frac{13}{32J}\cos 2\hat{\varphi} + \frac{r^2}{l^2}\frac{J\omega_o^2}{4}(\cos\hat{\varphi} - \cos 3\hat{\varphi}),
\end{aligned} \tag{C.10}$$

where we have chosen the normal-ordered form with respect to $\hat{p}_\varphi$ in order to expose the corrections to the classical expression (they are proportional to powers of $\hbar$). A symmetrized form displays the Hermitian character of $\hat{\mathcal{K}}^{(4)}$ more directly:

$$\begin{aligned}
\hat{\mathcal{K}}^{(4)} =& \frac{r^2}{l^2}\frac{3}{16J}\left(\hat{p}_\varphi^2(1-\cos 2\hat{\varphi}) + (1-\cos 2\hat{\varphi})\hat{p}_\varphi^2\right) \\
& - \hbar^2\frac{r^2}{l^2}\frac{13}{32J}\cos 2\hat{\varphi} + \frac{r^2}{l^2}\frac{J\omega_o^2}{4}(\cos\hat{\varphi} - \cos 3\hat{\varphi}).
\end{aligned} \tag{C.11}$$

Note that unlike the case of $\hat{\mathcal{K}}^{(2)}$, there is no quantization procedure that allows to get $\hat{\mathcal{K}}^{(4)}$ from $\mathcal{K}^{(4)}$ [41, 108, 159]. This is a direct consequence of Groenewold's theorem [67, 173]. Note that the calculation displayed in Eq. (C.10) can be easily automated with available computer algebra systems [110].

Finally, we also remark that in the limit $\omega_o \to 0$ this model provides a $\cos(2\hat{\varphi})$ potential that has received attention in the context of protected superconducting qubits [145].

Appendix D
Useful Identities

Without loss of generality, we assume $[\hat{a}, \hat{a}^\dagger] = \hbar$ so that all the expansions can be rewritten as series in $\hbar$. To recover the regular bosonic operators, set $\hbar = 1$. Normal ordering has been privileged due to its popularity in studying quantum onlinear dynamics. Formal proofs have been suppressed for the sake of brevity, but all of the identities below can be proven by induction, which we encourage the reader to do.

D.1 On Hilbert Space Operators

1.

$$(\hat{a}^\dagger + \hat{a})^m = \sum_{k=0}^{m} \binom{m}{k} \sum_{j=0}^{\min(k, m-k)} \frac{1}{2^j} j! \hbar^j \binom{k}{j} \binom{m-k}{j} \hat{a}^{\dagger k-j} \hat{a}^{m-k-j}. \tag{D.1}$$

2. We rewrite the above as

$$(\hat{a}^\dagger + \hat{a})^m = \sum_{j=0}^{\lfloor \frac{m}{2} \rfloor} \hbar^j \binom{m}{2j} (2j-1)!! \sum_{k=0}^{m-2j} \binom{m-2j}{k} \hat{a}^{\dagger k} \hat{a}^{m-2j-k}. \tag{D.2}$$

3. We find the normal ordered form of $\hat{a}^s \hat{a}^{\dagger m}$ for integers s and m to be

$$\hat{a}^s \hat{a}^{\dagger m} = \sum_{k=0}^{\min(s,m)} \hbar^k k! \binom{s}{k} \binom{m}{k} \hat{a}^{\dagger m-k} \hat{a}^{s-k}. \tag{D.3}$$

J. Venkatraman, *Controlling the Effective Hamiltonian of a Driven Quantum Superconducting Circuit*, Springer Theses,
https://doi.org/10.1007/978-3-031-83270-3

4. We find the normal ordered form of operator $\hat{O} = (\hat{a}^\dagger \hat{a})^n$ as

$$(\hat{a}^\dagger \hat{a})^n = \sum_{k=1}^{n} \hbar^{n-k} S(n,k) \hat{a}^{\dagger k} \hat{a}^k, \tag{D.4}$$

where $S(n,k)$ is the Sterling series of the second kind.

5.

$$\begin{aligned}
[\hat{a}^{\dagger r}\hat{a}^s, \hat{a}^{\dagger m}\hat{a}^n] &= \hat{a}^{\dagger r}\hat{a}^s\hat{a}^{\dagger m}\hat{a}^n - \hat{a}^{\dagger m}\hat{a}^n\hat{a}^{\dagger r}\hat{a}^s \\
&= \sum_{k=0}^{\min(s,m)} \hbar^k k! \binom{s}{k}\binom{m}{k} \hat{a}^{\dagger r+m-k}\hat{a}^{s+n-k} \\
&\quad - \sum_{k'=0}^{\min(n,r)} \hbar^{k'} k'! \binom{n}{k'}\binom{r}{k'} \hat{a}^{\dagger r+m-k'}\hat{a}^{s+n-k'}.
\end{aligned} \tag{D.5}$$

6. We know that $a|\alpha\rangle = \alpha|\alpha\rangle$. What about $a^\dagger|\alpha\rangle$? We evaluate it as

$$\begin{aligned}
a^\dagger|\alpha\rangle &= a^\dagger \mathcal{D}(\alpha)|0\rangle \\
&= \mathcal{D}(\alpha)\mathcal{D}^\dagger(\alpha)a^\dagger\mathcal{D}(\alpha)|0\rangle \\
&= \mathcal{D}(\alpha)(a^\dagger + \alpha^*)|0\rangle \\
&= \mathcal{D}(\alpha)|1\rangle + \alpha^*|\alpha\rangle,
\end{aligned} \tag{D.6}$$

where $\mathcal{D}(\alpha)$ corresponds to the displacement unitary operator defined as $e^{\alpha a^\dagger - \alpha^* a}$.

An excellent reference for more bosonic identities is [11]. Specific instances of the general identities that occur frequently throughout this dissertation work are displayed below for convenience.

$$\begin{aligned}
(\hat{a} + \hat{a}^\dagger)^2 &= \hat{a}^{\dagger 2} + 2\hat{a}^\dagger\hat{a} + \hat{a}^2 + \hbar \\
(\hat{a} + \hat{a}^\dagger)^3 &= \hat{a}^{\dagger 3} + 3\hat{a}^{\dagger 2}\hat{a} + 3\hat{a}^\dagger\hat{a}^2 + \hat{a}^3 + 3\hbar(\hat{a}^\dagger + \hat{a}) \\
(\hat{a}\hat{a}^\dagger)^2 &= \hat{a}^{\dagger 2}\hat{a}^2 + \hbar\hat{a}^\dagger\hat{a} \\
\hat{a}^2\hat{a}^{\dagger 2} - \hat{a}^{\dagger 2}\hat{a}^2 &= 4\hat{a}^\dagger\hat{a}\hbar + 2\hbar^2.
\end{aligned} \tag{D.7}$$

$$\begin{aligned}
[\hat{a}^2 + \hat{a}^{\dagger 2}, \hat{a}^\dagger\hat{a}] &= 2\hbar(a^2 - \hat{a}^{\dagger 2}) \\
[\hat{a}^2 + \hat{a}^{\dagger 2}, (\hat{a}^\dagger\hat{a})^2] &= 4\hbar(\hat{a}^\dagger\hat{a}^3 - \hat{a}^{\dagger 3}\hat{a}) + 4\hbar^2(\hat{a}^2 - \hat{a}^{\dagger 2}) \\
[\hat{a}^2 + \hat{a}^{\dagger 2}, \hat{a}^{\dagger 2}\hat{a}^2 - \hbar\hat{a}^\dagger\hat{a}] &= 4\hbar(\hat{a}^\dagger\hat{a}^3 - \hat{a}^{\dagger 3}\hat{a})
\end{aligned} \tag{D.8}$$

D.2 On Phase Space Functions

The $\star$-product between f and g in Fourier space [35]:

$$f \star g = \frac{1}{\hbar^2 \pi^2} \int dp' dp'' dx' dx'' f\left(x', p'\right) g\left(x'', p''\right) \times \exp\left(\frac{-2i}{\hbar}\left(p\left(x' - x''\right) + p'\left(x'' - x\right) + p''\left(x - x'\right)\right)\right), \tag{D.9}$$

see proof and an alternate form of Eq. (D.9) below.

Proof of Eq. (D.9)

We first write f and g as Fourier transforms as

$$f(x, p) = \int_{-\infty}^{\infty} dx' dp' \, f(x', p') \exp\left(-i2\pi(xp' - px')\right),$$

$$g(x, p) = \int_{-\infty}^{\infty} dx'' dp'' \, g(x'', p'') \exp\left(-i2\pi(xp'' - px'')\right).$$

We use the shift identity associated with the $\star$-product [35]:

$$f(x, p) \star g(x, p) = f\left(x + \frac{i\hbar}{2}\vec{\partial}_p, p - \frac{i\hbar}{2}\vec{\partial}_x\right) g(x, p), \tag{D.10}$$

and rewrite $f \star g$ as

$$f(x, p) \star g(x, p) = \int_{-\infty}^{\infty} dx' dp' dx'' dp'' \, f(x', p') g(x'', p'') \times \exp\left(-i2\pi(xp' - px')\right) \times \exp\left(\pi\hbar(p'\overrightarrow{\partial}_p + x'\overrightarrow{\partial}_x)\right) \times \exp\left(-i2\pi(xp'' - px'')\right). \tag{D.11}$$

Now, we simplify Eq. (D.11) using the following identities:

$$\exp(A\overrightarrow{\partial}_p) \exp(Bp) = \exp(AB) \times \exp(Bp)$$

$$\exp(C\overrightarrow{\partial}_x) \exp(Dx) = \exp(CD) \times \exp(Dx),$$

where in this case $A = \pi\hbar p'$, $B = i2\pi x''$, $C = \pi\hbar x'$, and $D = -i2\pi p''$. With this, and the substitutions $x' \to -x'\sqrt{\pi\hbar}$, $p' \to p'\sqrt{\pi\hbar}$, $x'' \to x''\sqrt{\pi\hbar}$, $p'' \to -p''\sqrt{\pi\hbar}$, Eq. (D.11) reduces to:

$$f(x,p) \star g(x,p) = \int_{-\infty}^{\infty} dx'dp'dx''dp''\, f(x',p')g(x'',p'') \times \exp\frac{-2i}{\hbar}\left(xp' - px' - p'x'' - x'p'' + xp'' - px''\right), \tag{D.12}$$

which when rearranged yields

$$f(x,p) \star g(x,p) = \int_{-\infty}^{\infty} dx'dp'dx''dp''\, f(x',p')g(x'',p'') \times \exp\frac{-2i}{\hbar}\left(p(x'-x'') + p'(x''-x) + p''(x-x')\right), \tag{D.13}$$

Note that by the following simple change of variables: $X' = x' - x$, $P' = p' - p$, $X'' = x'' - x$, $P'' = p'' - p$, we get an alternate form of Eq. (D.13):

$$f(x,p) \star g(x,p) = \int_{-\infty}^{\infty} dX'dP'dX''dP''\, f(X'+x, P'+p)g(X''+x, P''+p) \times \exp\frac{2i}{\hbar}\left(X'P'' - P'X''\right), \tag{D.14}$$

The expectation value of an observable $\hat{g}$ is given as

$$\langle \hat{g} \rangle = \mathrm{Tr}(\hat{\rho}\hat{g}) \tag{D.15}$$

$$= \int dxdp\, g(x,p) \star \rho(x,p) \tag{D.16}$$

$$= \int dxdp\, \rho(x,p) \star g(x,p) \tag{D.17}$$

$$= \int dxdp\, g(x,p)\rho(x,p). \tag{D.18}$$

The right hand side of Eq. (D.16) is obtained by taking a Weyl transform of the cyclic trace in Eq. (D.15). But the subsequent equalities in Eqs. (D.17) and (D.18) is by no means obvious; see proof in Appendix D.2. However, the elegant property that the expectation value is simply the integral of the regular (and not star) product allows one to employ the phase space formulation in quantum mechanics without knowing about the star product at all!

Proof of Eqs. (D.15)–(D.18)

We consider the expectation value of $\hat{g}$:

$$\langle \hat{g} \rangle = \mathrm{Tr}(\hat{\rho}\hat{g}) \tag{D.19a}$$

$$= \int_{-\infty}^{\infty} dxdp\, g(x,p) \star \rho(x,p) \tag{D.19b}$$

Using Eq. (D.13), we express Eq. (D.19b) as

$$\int_{-\infty}^{\infty} dxdp\, g(x,p) \star \rho(x,p)$$

$$= \int_{-\infty}^{\infty} dxdpdx'dp'dx''dp''\, g(x',p')\rho(x'',p'') \tag{D.19c}$$

$$\times \exp \frac{-2i}{\hbar} \left(p(x'-x'') + p'(x''-x) + p''(x-x')\right)$$

Observing that the integral over x and p do not act on g and ρ, we evaluate them in Eq. (D.19c) as:

$$\begin{aligned}\int_{-\infty}^{\infty} dxdp\, g(x,p) \star \rho(x,p) &= \int_{-\infty}^{\infty} dx'dp'dx''dp''\, g(x',p')\rho(x'',p'') \\ &\quad\times \exp \frac{-2i}{\hbar} \left(p'x'' - p''x'\right) \\ &\quad\times \int_{-\infty}^{\infty} dxdp\, \exp \frac{-2i}{\hbar} \left(p(x'-x'') - x(p'-p'')\right) \\ &= \int_{-\infty}^{\infty} dx'dp'dx''dp''\, g(x',p')\rho(x'',p'') \\ &\quad\times \exp \frac{-2i}{\hbar} \left(p'x'' - p''x'\right)\delta(x'-x'')\delta(p'-p'') \\ &= \int_{-\infty}^{\infty} dxdp\, g(x,p)\rho(x,p),\end{aligned} \tag{D.19d}$$

and this proves Eq. (D.18).

Appendix E
Tutorial on the Phase Space Formulation of Quantum Mechanics

A full quantum mechanical treatment can be developed in phase space without incurring in any semiclassical approximations [35, 67, 115]. For the sake of completeness, here we provide an overview on the mapping from operator-valued Hilbert space to quantum phase space and a few elemental techniques and identities. We focus here on Wigner phase space, and showcase that the Wigner transform is more than a visualization tool for states. We note that our treatment can be equivalently extended to other phase space formulations [35, 76, 124, 131, 170].

E.1 From Operator Hilbert Space to Wigner Phase Space (and Back)

The Wigner transform [163] of the density matrix $\hat{\rho}$ is the Wigner function $W(X, P)$, where X and P are standard phase space coordinates (not operators) with dimensions of position and momentum (see Appendix A for notation). We write this as

$$\mathfrak{W}\{\hat{\rho}\} = W(X, P).$$

Let us remind the reader of some crucial properties of the Wigner function. We have

$$\iint dXdP\, W(X, P) = 1, \tag{E.1}$$

where each integral runs from $-\infty$ to ∞ and we suppress the limits in the following text for simplicity. For a pure state, we further have

J. Venkatraman, *Controlling the Effective Hamiltonian of a Driven Quantum Superconducting Circuit*, Springer Theses,
https://doi.org/10.1007/978-3-031-83270-3

$$h \iint dX dP\, W(X, P)^2 = 1, \tag{E.2}$$

where $h = 2\pi\hbar$.

In general, we have

$$0 \le h^{n-1} \iint dX dP\, W(X, P)^n \le 1, \tag{E.3}$$

which corresponds to the positivity of the density matrix.

Likewise, for a generic operator $\hat{F}$, we introduce the phase space function $F(X, P) = \mathfrak{W}\{\hat{F}\}$.

In this framework, the average value of an Hermitian operator $\hat{F}$ can be written as

$$\langle \hat{F} \rangle = \iint dX dP\, F(X, P) W(X, P). \tag{E.4}$$

The transformation $\mathfrak{W}$ is invertible as appreciated by Groenewold [67]

$$\mathfrak{W}^{-1}\{W(X, P)\} = \hat{\rho}.$$

The inverse transformation $\mathfrak{W}^{-1}$ is know as the Weyl transformation [160].

In general, the Weyl transformation is

$$\hat{\rho} = \mathfrak{W}^{-1}\{W\} = \frac{1}{h} \iiiint dX dP dk dl\, W(X, P) e^{\frac{i}{\hbar}(k(\hat{X}-X)+l(\hat{P}-P))}, \tag{E.5}$$

where the characteristic function $C(l, k)$ defined as

$$C(l, k) = \iint dX dP\, e^{-\frac{i}{\hbar}(kX+lP)} W(X, P), \tag{E.6}$$

is the Fourier transform of the Wigner function and C is dimensionless.

Another useful formula is

$$W(X, P) = \frac{1}{h} \int dq\, e^{-iqP/\hbar} \langle X + q/2 | \hat{\rho} | X - q/2 \rangle, \tag{E.7}$$

where $\hat{\rho}$ is to be understood in the continuous position basis and therefore has the dimension of [1/position].

We now review simple operational rules to go from operator space to phase space functions and back without performing cumbersome integrals.

The Wigner and Weyl transformation take a particularly simple form for binomial expansions

$$\mathfrak{W}\{(\alpha\hat{X}+\beta\hat{P})^n\} = (\alpha X+\beta P)^n.$$

$$\mathfrak{W}^{-1}\{(\alpha X+\beta P)^n\} = (\alpha\hat{X}+\beta\hat{P})^n.$$

For non-symmetric expressions, the Wigner transform can be evaluated via a non-commutative Wigner phase space product, the celebrated Groenwold's star product.

E.2 An Introduction to the Star Product

We introduce the star product as

$$\mathfrak{W}(\hat{F}\hat{G}) = \mathfrak{W}(\hat{F}) \star \mathfrak{W}(\hat{G}) = F(X,P) \star G(X,P), \tag{E.8}$$

defined as (the exponential of the Poisson bracket):

$$\begin{aligned} F \star G &= \sum_{n=0}^{\infty}\sum_{k=0}^{n} \frac{(-1)^k}{n!}\left(\frac{i\hbar}{2}\right)^n \binom{n}{k} \partial_P^k \partial_X^{n-k} F \times \partial_P^{n-k}\partial_X^k G \\ &\equiv F \exp\left(\frac{i\hbar}{2}\left(\overleftarrow{\partial}_X \overrightarrow{\partial}_P - \overleftarrow{\partial}_P \overrightarrow{\partial}_X\right)\right) G \qquad \text{(E.9)} \\ &= FG + \frac{i\hbar}{2}\{F,G\} + \cdots \end{aligned}$$

Here $F\overleftarrow{\partial}_X G = (\partial_X F)G$ and $F\overrightarrow{\partial}_X G = F(\partial_X G)$, and we have introduced the Poisson bracket $\{F,G\} = \partial_X F \partial_P G - \partial_P F \partial_X G$. The star product can also be conveniently expressed in terms of complex-coordinates a and a^* as

$$F \star G \equiv F \exp\left(-\frac{1}{2}\left(\overleftarrow{\partial}_{a^*} \overrightarrow{\partial}_a - \overleftarrow{\partial}_a \overrightarrow{\partial}_{a^*}\right)\right) G.$$

It generalizes to a system of many particles (or many modes) as

$$F \star G = F \exp\left(\frac{i\hbar}{2}\sum_j \left(\overleftarrow{\partial}_{X_j} \overrightarrow{\partial}_{P_j} - \overleftarrow{\partial}_{P_j} \overrightarrow{\partial}_{X_j}\right)\right) G.$$

In Fourier space the star product becomes a phase factor: $\star \to e^{i\frac{\hbar}{2}(k_X k'_P - k'_X k_P)}$ [172]. This phase corresponds to an oriented area in reciprocal phase space. This is the simplest manifestation of the noncommutativity of the algebra of quantum mechanics in phase space.

Remarkably, the scalar product associated with the star product is the usual integral in phase space. For phase space functions in the Wigner representation F and G, we have

$$\iint dXdP\, F(X,P) \star G(X,P) = \iint dXdP\, F(X,P)G(X,P). \tag{E.10}$$

Note however that in general for any $F(X,P)$, $G(X,P)$, and $H(X,P)$,

$$\begin{aligned}&\iint dXdP\, F(X,P) \star G(X,P) \star H(X,P)\\ &\quad\neq \iint dXdP\, F(X,P)G(X,P)H(X,P).\end{aligned} \tag{E.11}$$

For non-symmetric expressions in $\hat{X}$ and $\hat{P}$, the above formulae can be employed to evaluate the Wigner transform. For example

$$\mathfrak{W}\{\hat{X}\hat{P}\hat{P}\} = XP^2 + i\hbar P$$

$$\mathfrak{W}\{\hat{P}\hat{P}\hat{X}\} = XP^2 - i\hbar P$$

$$\mathfrak{W}\{\hat{P}\hat{X}\hat{P}\} = XP^2.$$

We evaluate the Weyl transform of asymmetric expressions by symmetrizing it and replacing phase space functions by their corresponding operators. For example

$$\mathfrak{W}^{-1}\{XP^2\} = \frac{1}{3}(\hat{X}\hat{P}\hat{P} + \hat{P}\hat{X}\hat{P} + \hat{P}\hat{P}\hat{X}).$$

To find the Weyl transform of a high-degree polynomial of X and P, the Weyl-symmetrized form might be too tedious and McCoy [108] provided a shortcut to obtain polynomial expressions in the phase space representation. We review McCoy's formula in the next section.

E.3 The McCoy Formula for Obtaining Ordered Operators from Phase Space Functions

While a fully-symmetrized representation is usually inconvenient for polynomials of large degree, McCoy derived a set of formulae [108], each corresponding to a different representation of a Weyl transform. Here, we present two of them that yield operators that privilege the ordering of $\hat{X}$ (or $\hat{P}$).

Consider a phase space function $F(X, P)$. Its operator-valued correspondent $\hat{F}$ in normal order with respect to X is given by the McCoy formula [108] that reads:

$$\mathcal{F}(X, P) = e^{-i\frac{\hbar}{2}\partial_X\partial_P} F(X, P)$$

$$\mathcal{F}(X, P) = F(X, P) - \frac{i\hbar}{2}\partial_X\partial_P F(X, P) - \frac{1}{2!}\frac{\hbar^2}{2^2}\partial_X^2\partial_P^2 F(X, P) + \cdots$$

$$\hat{F} = (\mathcal{N}_X\mathcal{F})|_{(\hat{X},\hat{P})},$$

The functional (operator over real functions) $\mathcal{N}_X$ is carried out by writing its arguments with X factors (or P factors as indicated by the subindex of $\mathcal{N}$) to the left in each term and replace X, P with $\hat{X}$, $\hat{P}$ respectively. For example, if $F = XP$, we have $\mathcal{F} = XP - i\frac{\hbar}{2}$ which gives the correct and now ordered Hermitian expression for the operator $\hat{F} = \hat{X}\hat{P} - i\frac{\hbar}{2} = \frac{\hat{X}\hat{P}+\hat{P}\hat{X}}{2}$.

The inverse transform, is simply given $F(X, P) = e^{i\frac{\hbar}{2}\partial_X\partial_P}\mathcal{F}(X, P)$.

In terms of complex coordinates, $a = \frac{1}{\sqrt{2}}(x + ip)$ we adapt McCoy's formula [108]:

$$\mathcal{F}(a, a^*) = e^{\frac{1}{2}\partial_a\partial_{a^*}} F(a, a^*)$$

$$\hat{F} = (\mathcal{N}_{a^*}\mathcal{F})|_{(\hat{a},\hat{a}^\dagger)}$$

to get the normal ordered (with respect to a^*) result. For example, one has classically that $\frac{1}{2}(x^2 + p^2) = a^*a$. The correct quantization reads $F = aa^* \to \mathcal{F} = aa^* + 1/2 \to \hat{F} = \hat{a}^\dagger\hat{a} + 1/2$.

Application to Our Hamiltonian

If the Wigner phase space Kerr Hamiltonian reads $H = \Delta a^*a - Ka^{*2}a^2$ the corresponding operator is

$$F = a^{*2}a^2 \to \mathcal{F} = a^{*2}a^2 + 2a^*a + \frac{1}{2} \to \hat{F} = \hat{a}^{\dagger 2}\hat{a}^2 + 2\hat{a}^\dagger\hat{a} + \frac{1}{2},$$

$$\hat{H}/\hbar = (\Delta - 2K)\hat{a}^\dagger\hat{a} - K\hat{a}^{\dagger 2}\hat{a}^2,$$

where the oscillator frequency is renormalized by $2K$. This is the Lamb shift, and its origin is in the non commutativity of $\hat{a}$ and $\hat{a}^\dagger$, i.e. the vacuum fluctuations.

E.4 Groenewold's Theorem

Note that $\mathfrak{W}\left\{\frac{1}{i\hbar}[\hat{F}, \hat{G}]\right\} = \{\mathfrak{W}(\hat{F}), \mathfrak{W}(\hat{G})\} \neq \{\mathfrak{W}(\hat{F}), \mathfrak{W}(\hat{G})\}$. The quantum commutators do not correspond to the Poisson brackets: the theorem [67] states that such a mapping does not exist. We provide a practical consequence of the implications of this theorem to quantum Hamiltonian engineering in Appendix B of [153].

E.5 Dynamics of the Wigner Function: The Moyal Equation

The von-Neumann equation $\partial_t \hat{\rho} = \frac{1}{i\hbar}[\hat{H}, \hat{\rho}]$ (the density-operator version of the Schrödinger equation) transforms as

$$\partial_t W = \frac{1}{i\hbar}(H \star W - W \star H),$$

$$\partial_t W = \{H, W\}.$$

Here $H(X, P) = \mathfrak{W}(\hat{H})$ is the Hamiltonian function and we have introduced the Moyal bracket notation [115]. We refer the reader to [24] for a derivation of the equation of motion of the Wigner function from Schödinger's equation for the wavefunction without referring to the star product.

The exponential notation of the star product induces the name "Moyal sine bracket" since it can be written as

$$\partial_t W = H \frac{2}{\hbar} \sin\left(\frac{\hbar}{2}\left(\overleftarrow{\partial}_X \overrightarrow{\partial}_P - \overleftarrow{\partial}_P \overrightarrow{\partial}_X\right)\right) W.$$

Note that the Moyal equation is identical to Liouville equation plus quantum corrections coming from the expansion of the sine to higher orders of $\hbar$.

$$\partial_t W = \{H, W\} + \mathcal{O}(\hbar^2).$$

Interestingly, there is no corrections to $\mathcal{O}(\hbar)$. Importantly, the quantum corrections are proportional to $\hbar^2$ and to the nonlinear terms in the Hamiltonian. For quadratic Hamiltonians, all the quantum corrections vanish: the higher-order derivatives exterminate low-order polynomials (see the Appendix of [53]). Specifically, Gaussian transformations, i.e., those generated by quadratic Hamiltonians in the phase space coordinates, are classical in the sense that they are ruled by only the Poisson bracket. Thus, they would not develop negativities in the Wigner distribution if none would be present at the beginning.

E.6 Phase Space Formulation for Open Quantum Systems

So far, we have only discussed the phase space formulation for closed quantum systems. Indeed, one can extend the treatment to open systems as we demonstrate below. The Lindblad equation for single photon loss is given by

$$\partial_t \hat{\rho} = \frac{1}{i\hbar}[\hat{H}, \hat{\rho}] + \kappa \hat{a}\hat{\rho}\hat{a}^\dagger - \frac{\kappa}{2}(\hat{a}^\dagger \hat{a}\hat{\rho} + \hat{\rho}\hat{a}^\dagger \hat{a}). \tag{E.12}$$

Using Eqs. (E.8) and (E.9), we get the phase space formulation of Eq. (E.12) as

$$\mathfrak{W}\{\partial_t \hat{\rho}\} = \partial_t W,$$

$$\begin{aligned}\mathfrak{W}\left\{\frac{1}{i\hbar}[\hat{H}, \hat{\rho}]\right\} &= \{\mathfrak{W}(\hat{H}), \mathfrak{W}(\hat{\rho})\} \\ &= \{H, W\}\end{aligned}$$

$$\begin{aligned}\mathfrak{W}\{\hat{a}\hat{\rho}\hat{a}^\dagger\} &= a \star W \star a^* \\ &= aWa^* + \frac{1}{2}\partial_a W + \frac{1}{2}(\partial_{a^*}(Wa^*) + \frac{1}{2}\partial^2_{aa^*} W)\end{aligned}$$

$$\begin{aligned}\mathfrak{W}\{\hat{a}^\dagger \hat{a}\hat{\rho}\} &= \left(a^* a - \frac{1}{2}\right) \star W \\ &= aWa^* - \frac{1}{2}W + \frac{1}{2}(a^* \partial_{a^*} W - a\partial_a W) - \frac{1}{4}\partial^2_{aa^*} W\end{aligned}$$

$$\begin{aligned}\mathfrak{W}\{\hat{\rho}\hat{a}^\dagger \hat{a}\} &= W \star \left(a^* a - \frac{1}{2}\right) \\ &= aWa^* - \frac{1}{2}W - \frac{1}{2}(a^* \partial_{a^*} W - a\partial_a W) - \frac{1}{4}\partial^2_{aa^*} W\end{aligned}$$

Gathering all terms one directly gets

$$\partial_t W = \{H, W\} + \frac{\kappa}{2}\left(\partial^2_{aa^*} + \partial_a a + \partial_{a^*} a^*\right) W.$$

It is convenient to translate the above to x, p space

$$\partial_t W = \{H, W\} + \frac{\kappa}{2}\left(\partial_x^2 + \partial_p^2 + \partial_x x + \partial_p p\right) W. \tag{E.13}$$

By expressing the equation in x, p space in Eq. (E.13), the diffusion terms $\propto (\partial_x^2 + \partial_p^2)$ and the drag terms $\propto (\partial_x x + \partial_p p)$ associated to the fluctuation and the dissipation become evident. Note, that the Moyal sine bracket has only odd derivatives: the diffusion $(\partial_x^2 + \partial_p^2)$ cannot be canceled by Hamiltonian dynamics.

For finite temperature $\bar{n}_{\text{th}}$, the Lindblad master equation is

$$\partial_t \hat{\rho} = \frac{1}{i\hbar}[\hat{H}, \hat{\rho}] + \kappa(1 + \bar{n}_{\text{th}})\mathcal{D}[\hat{a}]\hat{\rho} + \kappa\bar{n}_{\text{th}}\mathcal{D}[\hat{a}^\dagger]\hat{\rho}, \tag{E.14}$$

where the dissipator $\mathcal{D}$ of the operator $\hat{O}$ is given by $\mathcal{D}[\hat{O}]\bullet := \hat{O}\bullet\hat{O}^\dagger - (\hat{O}^\dagger\hat{O}\bullet + \bullet\hat{O}^\dagger\hat{O})/2$.

It is straightforward to show that in the phase space formulation, Eq. (E.14) reads

$$\partial_t W = \{H, W\} + \frac{\kappa}{2}\left(\partial_a a + \partial_{a^*} a^*\right) W + \kappa\left(\frac{1}{2} + \bar{n}_{\text{th}}\right)\partial^2_{a^* a} W, \tag{E.15}$$

which reads in x, p space as

$$\partial_t W = \{H, W\} + \frac{\kappa}{2}\left(\partial_x x + \partial_p p\right) W + \frac{\kappa}{2}\left(\frac{1}{2} + \bar{n}_{\text{th}}\right)\left(\partial_x^2 + \partial_p^2\right) W. \tag{E.16}$$

Equation (E.16) is the quantum version of the Fokker-Planck equation, with the Poisson bracket replaced by the Moyal bracket and a quantum diffusion term corresponding to the zero point spread.

Note that for the Hamiltonian corresponding to Eq. (3.24), the solution for W from Eq. (E.16) will not yield the Boltzmann distribution in steady state, which perhaps is not surprising for an out-of-equilibrium driven problem [44].

Appendix F
Hamiltonian Transformation Under a Canonical Transformation

Claim Under a Lie transformation of the state $\rho \to \varrho = e^{L_S}\rho$, where $\partial_t \rho = \{H, \rho\}$, the transformed state ϱ is governed by the equation $\partial_t \varrho = \{K, \varrho\}$, where

$$K = e^{L_S} H + \int_0^1 d\epsilon\, e^{\epsilon L_S} \dot{S}. \tag{F.1}$$

Proof Let $\varrho = e^{L_S}\rho$, so that $\rho = e^{L_{-S}}\varrho$. We take the equation governing the time-evolution of ρ to be

$$\partial_t \rho = \{H, \rho\}, \tag{F.2a}$$

implying that

$$\partial_t (e^{L_{-S}}\varrho) = \{H, e^{L_{-S}}\varrho\} \tag{F.2b}$$

and using the product rule, we get

$$\partial_t \varrho = \{e^{L_S} H, \varrho\} - e^{L_S}\partial_t (e^{L_{-S}})\varrho, \tag{F.2c}$$

where we have used $e^{L_S}\{F, G\} = \{e^{L_S}F, e^{L_S}G\}$ [43].

Now, Eq. (B.4) is indeed true if

$$e^{L_S}\partial_t (e^{L_{-S}})\varrho = -L_{\int_0^1 d\epsilon\, e^{\epsilon L_S}\dot{S}}\varrho, \tag{F.3}$$

which we now prove. To evaluate the left hand side of Eq. (F.3), we invoke the following identity [62, 135, 141, 164]:

J. Venkatraman, *Controlling the Effective Hamiltonian of a Driven Quantum Superconducting Circuit*, Springer Theses,
https://doi.org/10.1007/978-3-031-83270-3

$$e^{-\square}\frac{d}{dt}e^{\square} = \left(\frac{1-e^{-\text{ad}_\square}}{\text{ad}_\square}\frac{d\square}{dt}\right), \tag{F.4a}$$

where the right hand side of Eq. (F.4a) can be further expanded using

$$\frac{1-e^{-\text{ad}_\square}}{\text{ad}_\square} = \sum_{k=0}^{\infty}\frac{(-1)^k}{(k+1)!}\text{ad}_\square^k.$$

In Eq. (F.4a), we have defined $\text{ad}_{\square_1}\square_2 = \lfloor\square_1, \square_2\rfloor = \square_1\square_2 - \square_2\square_1$, where $\lfloor\square_1, \square_2\rfloor$ represents the Lie bracket associated with the *operators* of the Lie algebra [62, 135, 141, 164]. Comparing Eq. (F.4a) with Eq. (F.3) and taking $\square \leftarrow (-L_S)$, Eq. (F.3) becomes

$$\begin{aligned} e^{L_S}\frac{\partial}{\partial t}e^{L_{-S}} &= -\left(\sum_{k=0}^{\infty}\frac{1}{(k+1)!}\text{ad}_{L_S}^k L_{\dot{S}}\right) \\ &= -\left(L_{\dot{S}} + \frac{1}{2!}\lfloor L_S, L_{\dot{S}}\rfloor + \frac{1}{3!}\lfloor L_S, \lfloor L_S, L_{\dot{S}}\rfloor\rfloor \right. \\ &\quad \left. + \cdots\right). \end{aligned} \tag{F.4b}$$

Note that we converted the total time-derivative in Eq. (F.4a) to a partial time-derivative in Eq. (F.4b) since we work in the active representation where the coordinates are static.

Now, we use an important property of the Lie bracket [20, 43]:

$$\begin{aligned} \lfloor L_F, L_G\rfloor &= L_{\{F,G\}} \\ &= L_{L_F G} \end{aligned} \tag{F.5}$$

to get

$$\lfloor L_F, \lfloor L_F, \lfloor \cdots \lfloor L_F, L_G\rfloor\rfloor\rfloor\rfloor = L_{L_F^n G}. \tag{F.6}$$

Equation (F.5) as written is simply the Jacobi identity and is generalized to Eq. (F.6) by induction. Employing Eq. (F.6) in Eq. (F.4b) at each order gives

$$\begin{aligned} e^{L_S}\frac{\partial}{\partial t}e^{L_{-S}} &= -\left(L_{\dot{S}} + L_{\frac{1}{2!}L_S\dot{S}} + L_{\frac{1}{3!}L_S^2\dot{S}} + \cdots\right) \\ &= -\left(L_{\sum_{n=0}^{\infty}\frac{1}{(n+1)!}L_S^n\dot{S}}\right) \\ &= -\left(L_{\int_0^1 d\epsilon\, e^{\epsilon L_S}\dot{S}}\right). \end{aligned} \tag{F.7}$$

This completes the proof.

Bibliography

1. Abanin DA, Papić Z (2017) Recent progress in many-body localization. Ann Phys 529(7):1700169
2. Acharya R, Aleiner I, Allen R, Andersen TI, Ansmann M, Arute F, Arya K, Asfaw A, Atalaya J, Babbush R, Bacon D, Bardin JC, Basso J, Bengtsson A, Boixo S, Bortoli G, Bourassa A, Bovaird J, Brill L, Broughton M, Buckley BB, Buell DA, Burger T, Burkett B, Bushnell N, Chen Y, Chen Z, Chiaro B, Cogan J, Collins R, Conner P, Courtney W, Crook AL, Curtin B, Debroy DM, Barba ADT, Demura S, Dunsworth A, Eppens D, Erickson C, Faoro L, Farhi E, Fatemi R, Burgos LF, Forati E, Fowler AG, Foxen B, Giang W, Gidney C, Gilboa D, Giustina M, Dau AG, Gross JA, Habegger S, Hamilton MC, Harrigan MP, Harrington SD, Higgott O, Hilton J, Hoffmann M, Hong S, Huang T, Huff A, Huggins WJ, Ioffe LB, Isakov SV, Iveland J, Jeffrey E, Jiang Z, Jones C, Juhas P, Kafri D, Kechedzhi K, Kelly J, Khattar T, Khezri M, Kieferová M, Kim S, Kitaev A, Klimov PV, Klots AR, Korotkov AN, Kostritsa F, Kreikebaum JM, Landhuis D, Laptev P, Lau K-M, Laws L, Lee J, Lee K, Lester BJ, Lill A, Liu W, Locharla A, Lucero E, Malone FD, Marshall J, Martin O, McClean JR, Mccourt T, McEwen M, Megrant A, Costa BM, Mi X, Miao KC, Mohseni M, Montazeri S, Morvan A, Mount E, Mruczkiewicz W, Naaman O, Neeley M, Neill C, Nersisyan A, Neven H, Newman M, Ng JH, Nguyen A, Nguyen M, Niu MY, O'Brien TE, Opremcak A, Platt J, Petukhov A, Potter R, Pryadko LP, Quintana C, Roushan P, Rubin NC, Saei N, Sank D, Sankaragomathi K, Satzinger KJ, Schurkus HF, Schuster C, Shearn MJ, Shorter A, Shvarts V, Skruzny J, Smelyanskiy V, Smith WC, Sterling G, Strain D, Szalay M, Torres A, Vidal G, Villalonga B, Heidweiller CV, White T, Xing C, Yao ZJ, Yeh P, Yoo J, Young G, Zalcman A, Zhang Y, Zhu N (2022) Suppressing quantum errors by scaling a surface code logical qubit. https://doi.org/10.1038/s41586-022-05434-1
3. Aidelsburger M, Atala M, Lohse M, Barreiro JT, Paredes B, Bloch I (2013) Realization of the Hofstadter Hamiltonian with ultracold atoms in optical lattices. Phys Rev Lett 111:185301
4. Albert VV, Jiang L (2014) Symmetries and conserved quantities in Lindblad master equations. Phys Rev A 89(2):022118
5. Alexandradinata A, Glazman L (2017) Geometric phase and orbital moment in quantization rules for magnetic breakdown. Phys Rev Lett 119:256601
6. Alexandradinata A, Glazman L (2018) Semiclassical theory of landau levels and magnetic breakdown in topological metals. Phys Rev B 97:144422
7. Altman E, Brown KR, Carleo G, Carr LD, Demler E, Chin C, DeMarco B, Economou SE, Eriksson MA, Fu K-MC, Greiner M, Hazzard KR, Hulet RG, Kollár AJ, Lev BL, Lukin MD, Ma R, Mi X, Misra S, Monroe C, Murch K, Nazario Z, Ni K-K, Potter AC, Roushan

J. Venkatraman, *Controlling the Effective Hamiltonian of a Driven Quantum Superconducting Circuit*, Springer Theses,
https://doi.org/10.1007/978-3-031-83270-3

P, Saffman M, Schleier-Smith M, Siddiqi I, Simmonds R, Singh M, Spielman I, Temme K, Weiss DS, Vučković J, Vuletić V, Ye J, Zwierlein M (2021) Quantum simulators: architectures and opportunities. PRX Quantum 2:017003
8. Axline CJ, Burkhart LD, Pfaff W, Zhang M, Chou K, Campagne-Ibarcq P, Reinhold P, Frunzio L, Girvin SM, Jiang L, Devoret MH, Schoelkopf RJ (2018) On-demand quantum state transfer and entanglement between remote microwave cavity memories. Nat Phys 14(7):705–710
9. Balantekin A, Takigawa N (1998) Quantum tunneling in nuclear fusion. Rev Mod Phys 70(1):77
10. Blais A, Grimsmo AL, Girvin SM, Wallraff A (2021) Circuit quantum electrodynamics. Rev Mod Phys 93(2):025005
11. Blasiak P, Horzela A, Penson KA, Solomon AI, Duchamp GH (2007) Combinatorics and Boson normal ordering: a gentle introduction. Am J Phys 75(7):639–646
12. Blencowe MP, Wang H (2020) Analogue gravity on a superconducting chip. Philos Trans R Soc A: Math Phys Eng Sci 378(2177):20190224
13. Bloch I, Dalibard J, Nascimbène S (2012) Quantum simulations with ultracold quantum gases. Nat Phys 8(4):267–276
14. Boettcher I, Bienias P, Belyansky R, Kollár AJ, Gorshkov AV (2020) Quantum simulation of hyperbolic space with circuit quantum electrodynamics: from graphs to geometry. Phys Rev A 102:032208
15. Bogoliubov NN, Mitropolsky YA (1961) Asymptotic methods in the theory of non-linear oscillations (in Russian) [English translation by Hindustan Pub., Delhi, 1961; Gordon and Breach, New York, 1961]
16. Breuer H-P, Petruccione F, et al (2002) The theory of open quantum systems. Oxford University Press on Demand, Oxford
17. Bukov M (2017) Floquet engineering in periodically driven closed quantum systems from dynamical localization to ultracold topological matter. PhD thesis, Boston University
18. Cage ME, Klitzing K, Chang A, Duncan F, Haldane M, Laughlin RB, Pruisken A, Thouless D (2012) The quantum Hall effect. Springer Science & Business Media, Berlin
19. Campagne-Ibarcq P, Eickbusch A, Touzard S, Zalys-Geller E, Frattini NE, Sivak VV, Reinhold P, Puri S, Shankar S, Schoelkopf RJ, et al (2020) Quantum error correction of a qubit encoded in grid states of an oscillator. Nature 584(7821):368–372
20. Cannas da Silva A (2001) Lectures on symplectic geometry, vol 1764. Lecture notes in mathematics. Springer, Berlin
21. Carmichael HJ (1999) Statistical methods in quantum optics 1: master equations and Fokker-Planck equations, vol 1. Springer Science & Business Media, Berlin
22. Carmichael HJ (2009) Statistical methods in quantum optics 2: non-classical fields. Springer Science & Business Media, Berlin
23. Casas F, Oteo JA, Ros J (2001) Floquet theory: exponential perturbative treatment. J Phys A Math Gen 34(16):3379–3388
24. Case WB (2008) Wigner functions and Weyl transforms for pedestrians. Am J Phys 76(10):937–946
25. Chamberland C, Noh K, Arrangoiz-Arriola P, Campbell ET, Hann CT, Iverson J, Putterman H, Bohdanowicz TC, Flammia ST, Keller A, Refael G, Preskill J, Jiang L, Safavi-Naeini AH, Painter O, Brandão FG (2022) Building a fault-tolerant quantum computer using concatenated cat codes. PRX Quantum 3:010329
26. Chapman BJ, de Graaf SJ, Xue SH, Zhang Y, Teoh J, Curtis JC, Tsunoda T, Eickbusch A, Read AP, Koottandavida A, et al (2022) A high on-off ratio beamsplitter interaction for gates on bosonically encoded qubits. arXiv preprint arXiv:2212.11929
27. Chávez-Carlos J, Lezama TLM, Cortiñas RG, Venkatraman J, Devoret MH, Batista VS, Pérez-Bernal F, Santos LF (2022) Spectral kissing and its dynamical consequences in the squeezed Kerr-nonlinear oscillator. https://doi.org/10.1038/s41534-023-00745-1
28. Chen Y-A, Nascimbène S, Aidelsburger M, Atala M, Trotzky S, Bloch I (2011) Controlling correlated tunneling and superexchange interactions with ac-driven optical lattices. Phys Rev Lett 107:210405

29. Chiorescu I, Nakamura Y, Harmans CM, Mooij J (2003) Coherent quantum dynamics of a superconducting flux qubit. Science 299(5614):1869–1871
30. Chono H, Kanao T, Goto H (2022) Two-qubit gate using conditional driving for highly detuned Kerr-nonlinear parametric oscillators. arXiv preprint arXiv:2204.03347
31. Clarke J, Braginski AI (2004) The SQUID handbook, vol 1. Wiley Online Library
32. Clerk AA, Devoret MH, Girvin SM, Marquardt F, Schoelkopf RJ (2010) Introduction to quantum noise, measurement, and amplification. Rev Mod Phys 82(2):1155
33. Cochrane PT, Milburn GJ, Munro WJ (1999) Macroscopically distinct quantum-superposition states as a bosonic code for amplitude damping. Phys Rev A 59(4):2631
34. Cohen J, Petrescu A, Shillito R, Blais A (2022) Reminiscence of classical chaos in driven transmons. https://doi.org/10.1103/PRXQuantum.4.020312
35. Curtright TL, Fairlie DB, Zachos CK (2013) A concise treatise on quantum mechanics in phase space. World Scientific Publishing Company, Singapore
36. del Pino J, Zilberberg O (2022) Dynamical gauge fields with bosonic codes. arXiv preprint arXiv:2211.12119
37. Deprit A (1969) Canonical transformations depending on a small parameter. Celest Mech 1(1):12–30
38. Devoret MH (2021) Does Brian Josephson's gauge-invariant phase difference live on a line or a circle? J Supercond Nov Magn 34(6):1633–1642
39. Devoret MH, Martinis JM, Clarke J (1985) Measurements of macroscopic quantum tunneling out of the zero-voltage state of a current-biased Josephson junction. Phys Rev Lett 55(18):1908
40. Devoret MH, et al (1995) Quantum fluctuations in electrical circuits. Les Houches, Session LXIII 7(8):133–135
41. Dirac PAM (1981) The principles of quantum mechanics. Number 27. Oxford University Press, Oxford
42. Doucet E, Reiter F, Ranzani L, Kamal A (2020) High fidelity dissipation engineering using parametric interactions. Phys Rev Res 2(2):023370
43. Dragt AJ, Finn JM (1976) Lie series and invariant functions for analytic symplectic maps. J Math Phys 17(12):2215–2227
44. Dykman M (2012) Fluctuating nonlinear oscillators: from nanomechanics to quantum superconducting circuits. Oxford University Press, Oxford
45. Eckardt A (2017) Colloquium: atomic quantum gases in periodically driven optical lattices. Rev Mod Phys 89(1):011004
46. Eckardt A, Anisimovas E (2015) High-frequency approximation for periodically driven quantum systems from a Floquet-space perspective. New J Phys 17(9):093039
47. Eddins A, Schreppler S, Toyli DM, Martin LS, Hacohen-Gourgy S, Govia LCG, Ribeiro H, Clerk AA, Siddiqi I (2018) Stroboscopic qubit measurement with squeezed illumination. Phys Rev Lett 120:040505
48. Fölling S, Trotzky S, Cheinet P, Feld M, Saers R, Widera A, Müller T, Bloch I (2007) Direct observation of second-order atom tunnelling. Nature 448(7157):1029–1032
49. Frattini NE (2021) Three-wave mixing in superconducting circuits: stabilizing cats with SNAILs. PhD thesis, Yale University
50. Frattini N, Vool U, Shankar S, Narla A, Sliwa K, Devoret M (2017) 3-Wave mixing Josephson dipole element. Appl Phys Lett 110(22):222603
51. Frattini NE, Sivak VV, Lingenfelter A, Shankar S, Devoret MH (2018) Optimizing the nonlinearity and dissipation of a SNAIL parametric amplifier for dynamic range. Phys Rev Appl 10(5):054020
52. Frattini NE, Sivak VV, Lingenfelter A, Shankar S, Devoret MH (2018) Optimizing the nonlinearity and dissipation of a SNAIL Parametric Amplifier for dynamic range. Phys Rev Appl 10:054020
53. Frattini NE, Cortiñas RG, Venkatraman J, Xiao X, Su Q, Lei CU, Chapman BJ, Joshi VR, Girvin S, Schoelkopf RJ, et al (2022) The squeezed Kerr oscillator: spectral kissing and phase-flip robustness. https://journals.aps.org/prx/abstract/10.1103/PhysRevX.14.031040

54. Friedman JR, Sarachik M, Tejada J, Ziolo R (1996) Macroscopic measurement of resonant magnetization tunneling in high-spin molecules. Phys Rev Lett 76(20):3830
55. Gao YY, Lester BJ, Zhang Y, Wang C, Rosenblum S, Frunzio L, Jiang L, Girvin SM, Schoelkopf RJ (2018) Programmable interference between two microwave quantum memories. Phys Rev X 8:021073
56. Gautier R, Sarlette A, Mirrahimi M (2022) Combined dissipative and Hamiltonian confinement of cat qubits. arXiv:2112.05545 [quant-ph] arXiv: 2112.05545
57. Girvin SM (2014) Circuit QED: superconducting qubits coupled to microwave photons. In: Quantum machines: measurement and control of engineered quantum systems. Oxford University Press, Oxford, pp 113–256
58. Goldman N, Dalibard J (2014) Periodically driven quantum systems: effective Hamiltonians and engineered gauge fields. Phys Rev X 4(3):031027
59. Goldstein H, Poole CP, Safko JL (2001) Classical mechanics (3rd edn.). Addison-Wesley. ISBN 978-0-201-65702-9
60. Goto H (2016) Universal quantum computation with a nonlinear oscillator network. Phys Rev A 93:050301
61. Goto H, Kanao T (2021) Chaos in coupled Kerr-nonlinear parametric oscillators. Phys Rev Res 3(4):043196
62. Grabowski J (1993) Derivative of the exponential mapping for infinite dimensional Lie groups. Ann Glob Anal Geom 11(3):213–220
63. Gravina L, Minganti F, Savona V (2022) A critical dissipative Schrödinger cat qubit. arXiv preprint arXiv:2208.04928
64. Griffiths DJ, Schroeter DF (2018) Introduction to quantum mechanics. Cambridge University Press, Cambridge
65. Grifoni M, Hänggi P (1998) Driven quantum tunneling. Phys Rep 304(5–6):229–354
66. Grimm A, Frattini NE, Puri S, Mundhada SO, Touzard S, Mirrahimi M, Girvin SM, Shankar S, Devoret MH (2020) Stabilization and operation of a Kerr-cat qubit. Nature 584(7820):205–209
67. Groenewold HJ (1946) On the principles of elementary quantum mechanics. In: On the principles of elementary quantum mechanics. Springer, Berlin, pp 1–56
68. Grossmann F, Dittrich T, Jung P, Hänggi P (1991) Coherent destruction of tunneling. Phys Rev Lett 67(4):516
69. Grozdanov T, Raković M (1988) Quantum system driven by rapidly varying periodic perturbation. Phys Rev A 38(4):1739
70. Gutzwiller MC (2013) Chaos in classical and quantum mechanics, vol 1. Springer Science & Business Media, Berlin
71. Habib S, Shizume K, Zurek WH (1998) Decoherence, chaos, and the correspondence principle. Phys Rev Lett 80:4361–4365
72. Haeberlen U, Waugh J (1968) Coherent averaging effects in magnetic resonance. Phys Rev 175(2):453
73. Haroche S (2013) Nobel lecture: controlling photons in a box and exploring the quantum to classical boundary. Rev Mod Phys 85:1083–1102
74. Haroche S, Raimond J-M (2006) Exploring the quantum: atoms, cavities, and photons. Oxford University Press, Oxford
75. Hensinger WK, Häffner H, Browaeys A, Heckenberg NR, Helmerson K, McKenzie C, Milburn GJ, Phillips WD, Rolston SL, Rubinsztein-Dunlop H, et al (2001) Dynamical tunnelling of ultracold atoms. Nature 412(6842):52–55
76. Hillery M, O'Connell RF, Scully MO, Wigner EP (1984) Distribution functions in physics: fundamentals. Phys Rep 106(3):121–167
77. Hsiao T-K, van Diepen C, Mukhopadhyay U, Reichl C, Wegscheider W, Vandersypen L (2020) Efficient orthogonal control of tunnel couplings in a quantum dot array. Phys Rev Appl 13:054018
78. Hu L, Ma Y, Cai W, Mu X, Xu Y, Wang W, Wu Y, Wang H, Song Y, Zou C-L, et al (2019) Quantum error correction and universal gate set operation on a binomial bosonic logical qubit. Nat Phys 15(5):503–508

79. Iachello F (1979) The interacting boson model. In: Neutron capture gamma-ray spectroscopy. Springer, Berlin, pp 23–35
80. Iachello F (1993) Dynamic symmetries and supersymmetries in nuclear physics. Rev Mod Phys 65(2):569
81. Iachello F (2006) Lie algebras and applications, vol 12. Springer, Berlin
82. Iachello F (2015) Lie algebras and Lie groups. In: Lie algebras and applications. Springer, Berlin, pp 53–56
83. Iachello F, Levine RD (1995) Algebraic theory of molecules. Oxford University Press, Oxford
84. Kanao T, Masuda S, Kawabata S, Goto H (2022) Quantum gate for a Kerr nonlinear parametric oscillator using effective excited states. Phys Rev Appl 18(1):014019
85. Kandala A, Mezzacapo A, Temme K, Takita M, Brink M, Chow JM, Gambetta JM (2017) Hardware-efficient variational quantum eigensolver for small molecules and quantum magnets. Nature 549(7671):242–246
86. Kapitza P (1951) Dynamic stability of the pendulum with vibrating suspension point. Sov Phys–JETP 21(5):588–597
87. Kaufman A, Lester B, Reynolds C, Wall M, Foss-Feig M, Hazzard K, Rey A, Regal C (2014) Two-particle quantum interference in tunnel-coupled optical tweezers. Science 345(6194):306–309
88. Kitaev A (2006) Anyons in an exactly solved model and beyond. Ann Phys 321(1):2–111. January Special Issue
89. Koch J, Terri MY, Gambetta J, Houck AA, Schuster DI, Majer J, Blais A, Devoret MH, Girvin SM, Schoelkopf RJ (2007) Charge-insensitive qubit design derived from the Cooper pair box. Phys Rev A 76(4):042319
90. Krantz P, Bengtsson A, Simoen M, Gustavsson S, Shumeiko V, Oliver W, Wilson C, Delsing P, Bylander J (2016) Single-shot read-out of a superconducting qubit using a Josephson parametric oscillator. Nat Commun 7(1):1–8
91. Krylov NM, Bogoliubov NN (1937) Introduction to non-linear mechanics (in Russian). Ac. of Sci. Ukr. SSR [English translation by Princeton University Press, Princeton, 1947]
92. Kurpiers P, Magnard P, Walter T, Royer B, Pechal M, Heinsoo J, Salathé Y, Akin A, Storz S, Besse J-C, Gasparinetti S, Blais A, Wallraff A (2018) Deterministic quantum state transfer and remote entanglement using microwave photons. Nature 558(7709):264–267
93. Kwon S, Watabe S, Tsai J-S (2022) Autonomous quantum error correction in a four-photon Kerr parametric oscillator. npj Quantum Inf 8(1):1–8
94. Landau LD, Lifshitz EM (1976) Mechanics: Volume 1, vol 1. Butterworth-Heinemann, Oxford
95. Landau LD, Lifshitz E (1991) Course of theoretical physics, quantum mechanics non-relativistic theory. Pergamon, Oxford
96. Leghtas Z, Touzard S, Pop IM, Kou A, Vlastakis B, Petrenko A, Sliwa KM, Narla A, Shankar S, Hatridge MJ, et al (2015) Confining the state of light to a quantum manifold by engineered two-photon loss. Science 347(6224):853–857
97. Lescanne R, Villiers M, Peronnin T, Sarlette A, Delbecq M, Huard B, Kontos T, Mirrahimi M, Leghtas Z (2020) Exponential suppression of bit-flips in a qubit encoded in an oscillator. Nat Phys 16(5):509–513
98. Lignier H, Sias C, Ciampini D, Singh Y, Zenesini A, Morsch O, Arimondo E (2007) Dynamical control of matter-wave tunneling in periodic potentials. Phys Rev Lett 99(22):220403
99. Lörch N, Zhang Y, Bruder C, Dykman MI (2019) Quantum state preparation for coupled period tripling oscillators. Phys Rev Res 1(2)
100. Loss D, DiVincenzo DP, Grinstein G (1992) Suppression of tunneling by interference in half-integer-spin particles. Phys Rev Lett 69(22):3232
101. Manucharyan VE, Koch J, Glazman LI, Devoret MH (2009) Fluxonium: single cooper-pair circuit free of charge offsets. Science 326(5949):113–116
102. Marthaler M, Dykman M (2006) Switching via quantum activation: a parametrically modulated oscillator. Phys Rev A 73(4):042108

103. Marthaler M, Dykman M (2007) Quantum interference in the classically forbidden region: a parametric oscillator. Phys Rev A 76(1):010102
104. Martinez EA, Muschik CA, Schindler P, Nigg D, Erhard A, Heyl M, Hauke P, Dalmonte M, Monz T, Zoller P, Blatt R (2016) Real-time dynamics of lattice gauge theories with a few-qubit quantum computer. Nature 534(7608):516–519
105. Masluk NA (2013) Reducing the losses of the fluxonium artificial atom. Yale University
106. Mattuck RD (1992) A guide to Feynman diagrams in the many-body problem. Courier Corporation, North Chelmsford
107. Matveev K, Larkin A, Glazman L (2002) Persistent current in superconducting nanorings. Phys Rev Lett 89(9):096802
108. McCoy NH (1932) On the function in quantum mechanics which corresponds to a given function in classical mechanics. Proc Natl Acad Sci 18(11):674–676
109. Merzbacher E (2002) The early history of quantum tunneling. Phys Today 55(8):44–49
110. Meurer A, Smith CP, Paprocki M, Čertík O, Kirpichev SB, Rocklin M, Kumar A, Ivanov S, Moore JK, Singh S, et al (2017) Sympy: symbolic computing in python. PeerJ Comput Sci 3:e103
111. Mikami T, Kitamura S, Yasuda K, Tsuji N, Oka T, Aoki H (2016) Brillouin-Wigner theory for high-frequency expansion in periodically driven systems: application to Floquet topological insulators. Phys Rev B 93(14):144307
112. Minev ZK, Leghtas Z, Mundhada SO, Christakis L, Pop IM, Devoret MH (2021) Energy-participation quantization of Josephson circuits. npj Quantum Inf 7(1):1–11
113. Mirrahimi M, Rouchon P (2015) Dynamics and control of open quantum systems (unpublished lecture notes). https://cas.minesparis.psl.eu/~rouchon/MasterUPMC/LectureNotes-03-18.pdf
114. Mirrahimi M, Leghtas Z, Albert VV, Touzard S, Schoelkopf RJ, Jiang L, Devoret MH (2014) Dynamically protected cat-qubits: a new paradigm for universal quantum computation. New J Phys 16(4):045014
115. Moyal JE (1949) Quantum mechanics as a statistical theory. In: Mathematical proceedings of the Cambridge Philosophical Society, vol 45. Cambridge University Press, Cambridge, pp 99–124
116. Nayfeh AH (2008) Perturbation methods. Wiley, New York
117. Nigg SE, Paik H, Vlastakis B, Kirchmair G, Shankar S, Frunzio L, Devoret MH, Schoelkopf RJ, Girvin SM (2012) Black-Box superconducting circuit quantization. Phys Rev Lett 108(24):240502
118. Noguchi A, Shikano Y, Toyoda K, Urabe S (2014) Aharonov–bohm effect in the tunnelling of a quantum rotor in a linear Paul trap. Nat Commun 5(1):1–6
119. Ofek N, Petrenko A, Heeres R, Reinhold P, Leghtas Z, Vlastakis B, Liu Y, Frunzio L, Girvin S, Jiang L, et al (2016) Extending the lifetime of a quantum bit with error correction in superconducting circuits. Nature 536(7617):441–445
120. Ong F, Boissonneault M, Mallet F, Doherty A, Blais A, Vion D, Esteve D, Bertet P (2013) Quantum heating of a nonlinear resonator probed by a superconducting qubit. Phys Rev Lett 110(4):047001
121. Petrescu A, Malekakhlagh M, Türeci HE (2020) Lifetime renormalization of driven weakly anharmonic superconducting qubits. II. The readout problem. Phys Rev B 101(13):134510
122. Place AP, Rodgers LV, Mundada P, Smitham BM, Fitzpatrick M, Leng Z, Premkumar A, Bryon J, Vrajitoarea A, Sussman S, et al (2021) New material platform for superconducting transmon qubits with coherence times exceeding 0.3 milliseconds. Nat Commun 12(1):1–6
123. Pop IM, Geerlings K, Catelani G, Schoelkopf RJ, Glazman LI, Devoret MH (2014) Coherent suppression of electromagnetic dissipation due to superconducting quasiparticles. Nature 508(7496):369–372
124. Puri RR, et al (2001) Mathematical methods of quantum optics, vol 79. Springer, Berlin
125. Puri S, Andersen CK, Grimsmo AL, Blais A (2017) Quantum annealing with all-to-all connected nonlinear oscillators. Nat Commun 8(1):1–9

126. Puri S, Boutin S, Blais A (2017) Engineering the quantum states of light in a Kerr-nonlinear resonator by two-photon driving. npj Quantum Inf 3(1):1–7
127. Puri S, Grimm A, Campagne-Ibarcq P, Eickbusch A, Noh K, Roberts G, Jiang L, Mirrahimi M, Devoret MH, Girvin SM (2019) Stabilized cat in a driven nonlinear cavity: a fault-tolerant error syndrome detector. Phys Rev X 9:041009
128. Puri S, St-Jean L, Gross JA, Grimm A, Frattini NE, Iyer PS, Krishna A, Touzard S, Jiang L, Blais A, Flammia ST, Girvin SM (2020) Bias-preserving gates with stabilized cat qubits. Sci Adv 6(34):eaay5901
129. Putterman H, Iverson J, Xu Q, Jiang L, Painter O, Brandão FG, Noh K (2022) Stabilizing a Bosonic qubit using colored dissipation. Phys Rev Lett 128(11):110502
130. Rahav S, Gilary I, Fishman S (2003) Effective Hamiltonians for periodically driven systems. Phys Rev A 68:013820
131. Raimond J-M, Haroche S (2006) Exploring the quantum, vol 82. Oxford University Press, Oxford, p 86
132. Ramos R, Spierings D, Racicot I, Steinberg AM (2020) Measurement of the time spent by a tunnelling atom within the barrier region. Nature 583(7817):529–532
133. Roberts D, Clerk AA (2020) Driven-dissipative quantum Kerr resonators: new exact solutions, photon blockade and quantum bistability. Phys Rev X 10(2):021022
134. Rosenblum S, Gao YY, Reinhold P, Wang C, Axline CJ, Frunzio L, Girvin SM, Jiang L, Mirrahimi M, Devoret MH, Schoelkopf RJ (2018) A cnot gate between multiphoton qubits encoded in two cavities. Nat Commun 9(1):652
135. Rossmann W (2006) Lie groups: an introduction through linear groups, vol 5. Oxford University Press on Demand, Oxford
136. Royer B, Singh S, Girvin SM (2020) Stabilization of finite-energy Gottesman-Kitaev-preskill states. Phys Rev Lett 125:260509
137. Ruiz D, Gautier R, Guillaud J, Mirrahimi M (2022) Squeezed Kerr quantum oscillator with multiple spectral degeneracies. https://doi.org/10.1103/PhysRevA.107.042407
138. Sank D, Chen Z, Khezri M, Kelly J, Barends R, Campbell B, Chen Y, Chiaro B, Dunsworth A, Fowler A, et al (2016) Measurement-induced state transitions in a superconducting qubit: beyond the rotating wave approximation. Phys Rev Lett 117(19):190503
139. Schoelkopf RJ, Clerk AA, Girvin SM, Lehnert KW, Devoret MH (2003) Qubits as spectrometers of quantum noise. Springer Netherlands, Dordrecht, pp 175–203
140. Schreier JA, Houck AA, Koch J, Schuster DI, Johnson BR, Chow JM, Gambetta JM, Majer J, Frunzio L, Devoret MH, et al (2008) Suppressing charge noise decoherence in superconducting charge qubits. Phys Rev B 77(18):180502
141. Schur F (1891) Zur theorie der endlichen transformationsgruppen. Math Ann 38(2):263–286
142. Shillito R, Petrescu A, Cohen J, Beall J, Hauru M, Ganahl M, Lewis AGM, Vidal G, Blais A (2022) Dynamics of transmon ionization. https://doi.org/10.1103/PhysRevApplied.18.034031
143. Shirley JH (1965) Solution of the Schrödinger equation with a hamiltonian periodic in time. Phys Rev 138(4B):B979
144. Sivak V, Frattini N, Joshi V, Lingenfelter A, Shankar S, Devoret M (2019) Kerr-free three-wave mixing in superconducting quantum circuits. Phys Rev Appl 11(5):054060
145. Smith WC, Kou A, Xiao X, Vool U, Devoret MH (2020) Superconducting circuit protected by two-Cooper-pair tunneling. npj Quantum Inf 6(1):8
146. Stein W, et al (2021) Sage mathematics software (Version 8.6.0). The Sage Development Team. http://www.sagemath.org
147. Sturge M (1968) The Jahn-Teller effect in solids. In: Solid state physics, vol 20. Elsevier, Amsterdam, pp 91–211
148. Suzuki Y, Kawabata S, Yamamoto T, Masuda S (2022) Quantum state tomography for Kerr parametric oscillators. arXiv preprint arXiv:2212.14627
149. Tomsovic S, Ullmo D (1994) Chaos-assisted tunneling. Phys Rev E 50(1):145
150. Touzard S, Kou A, Frattini NE, Sivak VV, Puri S, Grimm A, Frunzio L, Shankar S, Devoret MH (2019) Gated conditional displacement readout of superconducting qubits. Phys Rev Lett 122:080502

151. Unruh WG (1976) Notes on black-hole evaporation. Phys Rev D 14:870–892
152. Venkatraman J, Xiao X, Cortiñas RG, Devoret MH (2022) On the static effective Lindbladian of the squeezed Kerr oscillator. https://doi.org/10.1103/PhysRevA.110.042411
153. Venkatraman J, Xiao X, Cortiñas RG, Eickbusch A, Devoret MH (2022) Static effective Hamiltonian of a rapidly driven nonlinear system. Phys Rev Lett 129(10):100601
154. Venkatraman J, Cortiñas RG, Frattini NE, Xiao X, Devoret MH (2024) A driven kerr oscillator with two-fold degeneracies for qubit protection. Proc Natl Acad Sci 121(24):e2311241121
155. Vijay R, Devoret M, Siddiqi I (2009) Invited review article: the Josephson bifurcation amplifier. Rev Sci Instrum 80(11):111101
156. Wang C, Gao YY, Reinhold P, Heeres RW, Ofek N, Chou K, Axline C, Reagor M, Blumoff J, Sliwa KM, Frunzio L, Girvin SM, Jiang L, Mirrahimi M, Devoret MH, Schoelkopf RJ (2016) A Schrödinger cat living in two boxes. Science 352(6289):1087–1091
157. Wang CS, Curtis JC, Lester BJ, Zhang Y, Gao YY, Freeze J, Batista VS, Vaccaro PH, Chuang IL, Frunzio L, Jiang L, Girvin SM, Schoelkopf RJ (2020) Efficient multiphoton sampling of molecular vibronic spectra on a superconducting bosonic processor. Phys Rev X 10:021060
158. Wernsdorfer W, Sessoli R (1999) Quantum phase interference and parity effects in magnetic molecular clusters. Science 284(5411):133–135
159. Weyl H (1927) Quantenmechanik und gruppentheorie. Z Phys 46(1):1–46
160. Weyl H (1931) The theory of groups and quantum mechanics. E. P. Dutton & Co., New York
161. Wielinga B, Milburn GJ (1993) Quantum tunneling in a Kerr medium with parametric pumping. Phys Rev A 48:2494–2496
162. Wielinga B, Milburn GJ (1993) Quantum tunneling in a kerr medium with parametric pumping. Phys Rev A 48:2494–2496
163. Wigner E (1932) On the quantum correction for thermodynamic equilibrium. Phys Rev 40:749–759
164. Wilcox RM (1967). Exponential operators and parameter differentiation in quantum physics. J Math Phys 8(4):962–982
165. Wilson CM, Johansson G, Pourkabirian A, Simoen M, Johansson JR, Duty T, Nori F, Delsing P (2011) Observation of the dynamical casimir effect in a superconducting circuit. Nature 479(7373):376–379
166. Wintersperger K, Braun C, Ünal FN, Eckardt A, Liberto MD, Goldman N, Bloch I, Aidelsburger M (2020) Realization of an anomalous Floquet topological system with ultracold atoms. Nat Phys 16(10):1058–1063
167. Wustmann W, Shumeiko V (2013) Parametric resonance in tunable superconducting cavities. Phys Rev B 87(18):184501
168. Wustmann W, Shumeiko V (2019) Parametric effects in circuit quantum electrodynamics. Low Temp Phys 45(8):848–869
169. Xiao Z, Doucet E, Noh T, Ranzani L, Simmonds R, Govia L, Kamal A (2021) Perturbative diagonalization for time-dependent strong interactions. arXiv preprint arXiv:2103.09260
170. Xiao X, Venkatraman J, Cortinas RG, Chowdhury S, Devoret MH (2023) A diagrammatic method to compute the effective hamiltonian of driven nonlinear oscillators. arXiv preprint arXiv:2304.13656
171. Yan F, Gustavsson S, Kamal A, Birenbaum J, Sears AP, Hover D, Gudmundsen TJ, Rosenberg D, Samach G, Weber S, et al (2016) The flux qubit revisited to enhance coherence and reproducibility. Nat Commun 7(1):1–9
172. Zachos C (2000) Geometrical evaluation of star products. J Math Phys 41(7):5129–5134
173. Zachos C, Fairlie D, Curtright T (2005) Quantum mechanics in phase space: an overview with selected papers
174. Zeuch D, Hassler F, Slim JJ, DiVincenzo DP (2020) Exact rotating wave approximation. Ann Phys 423:168327
175. Zhang Y, Dykman MI (2019) Nonlocal random walk over Floquet states of a dissipative nonlinear oscillator. Phys Rev E 100:052148
176. Zhang Y, Gosner J, Girvin SM, Ankerhold J, Dykman MI (2017) Time-translation-symmetry breaking in a driven oscillator: from the quantum coherent to the incoherent regime. Phys Rev A 96:052124

177. Zhang Y, Lester BJ, Gao YY, Jiang L, Schoelkopf RJ, Girvin SM (2019) Engineering bilinear mode coupling in circuit qed: theory and experiment. Phys Rev A 99:012314
178. Zorin AB, Makhlin Y (2011) Period-doubling bifurcation readout for a Josephson qubit. Phys Rev B 83(22):224506
179. Zurek WH (2003) Decoherence and the transition from quantum to classical – revisited
180. Zurek WH, Paz JP (1994) Decoherence, chaos, and the second law. Phys Rev Lett 72:2508–2511
181. Zurek WH, Paz JP (1995) Zurek and Paz reply. Phys Rev Lett 75:351–351
182. Zurek WH, Paz JP (1999) Why we don't need quantum planetary dynamics: decoherence and the correspondence principle for chaotic systems. In: Epistemological and experimental perspectives on quantum physics. Springer, Berlin, pp 167–177

The manufacturer's authorised representative in the EU is Springer Nature Customer Service Centre GmbH, Europaplatz 3, 69115 Heidelberg, Germany. If you have any concerns regarding our products, please contact ProductSafety@springernature.com

Printed and bound by CPI Group (UK) Ltd, Croydon, CR0 4YY
30/09/2026
02225607-0009